Fatouma Mohamed Abdoul-Latif
Ayoub Ainane
Tarik Ainane

The Mysteries of Khat: Hallucinations or Therapy?

Fatouma Mohamed Abdoul-Latif
Ayoub Ainane
Tarik Ainane

The Mysteries of Khat: Hallucinations or Therapy?

ScienciaScripts

Imprint

Any brand names and product names mentioned in this book are subject to trademark, brand or patent protection and are trademarks or registered trademarks of their respective holders. The use of brand names, product names, common names, trade names, product descriptions etc. even without a particular marking in this work is in no way to be construed to mean that such names may be regarded as unrestricted in respect of trademark and brand protection legislation and could thus be used by anyone.

Cover image: www.ingimage.com

This book is a translation from the original published under ISBN 978-620-6-72143-7.

Publisher:
Sciencia Scripts
is a trademark of
Dodo Books Indian Ocean Ltd. and OmniScriptum S.R.L publishing group

120 High Road, East Finchley, London, N2 9ED, United Kingdom
Str. Armeneasca 28/1, office 1, Chisinau MD-2012, Republic of Moldova, Europe
Printed at: see last page
ISBN: 978-620-8-08353-3

"The Mysteries of Khat: Hallucinations or Therapy?"

Authors :

Dr Fatouma Mohamed Abdoul-Latif

Dr Ayoub Ainane

Dr Tarik Ainane

Contents :

Book summary:

Khat (*Catha edulis* Forsk) is a plant native to the Arabian Peninsula and East Africa, deeply rooted in the social and cultural traditions of these regions. Its chewing is a common practice in Djibouti and Yemen, seen as a way of preserving food traditions and adapting to local environments. Khat's psychoactive effects are mainly due to cathinone, a compound similar to amphetamines, as well as other phenylethylamines. The main effects of khat are euphoric and stimulating, enhancing the physical and mental capacities of users. However, these effects are accompanied by negative side effects, including health risks such as palpitations, tachycardia, vomiting and headaches. Claimed benefits of khat include reduced blood sugar levels, relief from asthma and gastrointestinal disorders.

Legislation on khat varies considerably around the world. In France, it has been listed as a narcotic since 1957. In Yemen, it has a significant impact on social and economic life. In Europe, notably the UK, khat importation is influenced by migrants from regions where it is commonly consumed. Synthetic drugs derived from khat, such as mephedrone, are gaining in popularity due to their low cost and the declining quality of other drugs.

Research has explored the use of khat alkaloids for pharmaceutical treatments. However, many of these treatments have become obsolete, with the exception of a smoking cessation aid. The recently-discovered cytotoxic properties of khat are arousing interest in its potential as a chemotherapeutic agent.

Despite the potential benefits, this book highlights several points, including the effects of khat, the variability of reported effects and the methodological limitations of some studies. Furthermore, the disparity in the legality and regulation of khat across different countries complicates the establishment of coherent and effective public health policies.

In conclusion, khat is a multifaceted plant, rooted in tradition but posing modern public health and regulatory challenges. Its therapeutic potential continues to be explored, despite the many obstacles and side effects associated with its consumption.

Introduction

Catha edulis Forsk, commonly known as khat, is a plant native to the Arabian Peninsula and parts of East Africa, belonging to the Celastraceae family. Its cultivation is deeply rooted in the ancient traditions of these regions, where khat chewing is seen as a way of preserving food traditions while adapting to local environments.

In Sana'a, the capital of Yemen, bustling markets offer small bundles of bare branches, wrapped in cloth or plastic to preserve the freshness of the oval khat leaves. Although the exact origin of the plant is debated, its presence in Yemeni culture has been undeniable for centuries. According to one local official, "there is no event in Yemen without khat". This tradition is not isolated, as khat also plays a significant role in Djibouti, where it is imported mainly from Ethiopia and consumed regularly by a large proportion of the population. In Djibouti, the practice of chewing khat is deeply rooted in local traditions and presents particular challenges in terms of public health and economic management.

Khat is known for its psychoactive effects, mainly due to cathinone, a compound structurally similar to amphetamines. In addition to cathinone, khat leaves contain other phenylethylamine compounds, such as cathine and beta-cathinone, as well as tannins, vitamins, minerals and flavonoids in significant proportions. These substances give khat sympathomimetic and euphoric effects, stimulating the physical and mental capacities of consumers. However, khat has both positive and negative effects. Its advocates point to benefits such as reduced blood sugar levels, its use as a remedy for asthma and its effectiveness in alleviating gastrointestinal disorders. On the other hand, drawbacks emerge: khat consumption can entail health risks, affect various aspects of life, and have medical and socio-economic repercussions.

On the scientific front, although research into the properties of khat is limited, some researchers are exploring its potential therapeutic applications for a variety of ailments. However, there are warnings about its harmful effects on the cardiovascular system, the central nervous system and mental health.

No definitive evidence has yet been established for a specific mechanism of toxicity associated with khat. Studies focus mainly on toxicity associated with cathine and other phytopharmaceutical compounds. It is also notable that khat possesses antioxidant properties which, although traditionally beneficial, may present risks at high concentrations. To fully understand the issues surrounding khat, it is essential to explore its cultural and historical roots, as well as consumption patterns. We need to examine how khat is grown and harvested, and determine its chemical, pharmacological and pharmacokinetic aspects. The impact of khat consumption on society and the economy is also crucial. As an active substance, khat poses significant challenges in terms of public health and legal regulation, aspects which we will detail in depth.

PART I
GENERAL INFORMATION ON KHAT

1. History of Khat

The precise origins of khat remain unclear. However, it is generally accepted that its consumption was widespread in Ethiopia as early as the XVème century. The practice then spread to the south-western Arabian Peninsula. Early Arabic manuscripts indicate that khat was already present in Yemen at the time of the Ethiopian invasion in the XVIème century.

Khat was probably introduced to Yemen from Turkistan in the late Xème century AD. According to some sources, it was brought from Ethiopia around 1345 AD. At the time, it was used to treat biliary diseases and soothe the stomach and liver. An Arabic medical compendium, *Kitab al-Saidana fi al-Tibb* by Al-Biruni, already mentions khat leaves as a remedy for depression, and the plant was also reputed for its aphrodisiac properties.

The first Western report on khat dates back to the 18th^{ème} century, when the Finnish botanist Pehr Forskal identified it in 1765 and named it *Catha edulis*. Unfortunately, Forskal did not live long enough to publish his research. His work was published in 1775 by Carsten Niebuhr, the sole survivor of the first European scientific expedition to Arabia. In his observations, Niebuhr notes that "the Arabs do not use opium like the Turks and Persians; instead, they chew khat. These are the buds of a certain plant, transported in small boxes from the hills of Yemen."

In homage to Pehr Forskal, Carsten Niebuhr called khat *Catha edulis* (Vahl) Forssk. ex Endl. Nevertheless, other names were attributed to the plant by various travelers in the XIXème century who crossed Arabia and East Africa.

2. Khat Habits and Consumption Patterns

Khat is used ritually in Yemen, Ethiopia, Djibouti, Somalia and Kenya, mainly for its stimulating effects. It is consumed by chewing the fresh young leaves, as well as the tender shoots at the ends of the plant. The consumption method is unique: the consumer regularly adds new leaves to the current chew. This is stored in the cheek and chewed slowly to extract all its juices. The residue is then spit out or swallowed.

Due to the bitterness of khat, chewing is often accompanied by drinks such as cool water, tea, coffee or Coca-Cola. During consumption sessions, which can last up to four hours, it is common for consumers to smoke cigarettes or use a hookah. Each session can involve the consumption of 100 to 200 grams of fresh leaves, equivalent to an oral dose of 5 mg of amphetamines.

Photo 1a: Khat session

Khat has long been used in a similar way to coca in the Andes of South America. In rural areas, it is traditionally chewed to relieve the fatigue associated with agricultural work. It also plays a role in religious celebrations and family gatherings, such as weddings and circumcisions. Drivers, merchants and students use it for its energizing effects, helping them to stay awake. What's more, khat promotes social interaction thanks to the euphoria it procures and the time spent chewing.

The quality of khat leaves also influences the way they are consumed. The large leaves, too hard to chew, are often dried and used to prepare infusions known as Abyssinian tea or Somali tea, depending on the region. The dried leaves can also be smoked like tobacco. Another preparation involves mixing the dried leaves with water, honey and spices to create a paste that is chewed and swallowed in a manner similar to traditional consumption.

In other contexts, khat wood is used in southern Africa for furniture and construction because of its hardness and resistance to termites. It can also be transformed into paper pulp, serving as an excellent blotting paper. The wood is also used to make handles for agricultural tools and kitchen utensils such as pots and spoons.

Photo 1b: Khat session

3. Khat Distribution Area

Catha edulis, commonly known as khat, is a plant whose natural range encompasses mainly the Horn of Africa, southeast Africa and the southern Arabian Peninsula. More specifically, it is found in the following regions:
- ✓ **In Africa:** Djibouti, Ethiopia, Somalia, Kenya, Tanzania, Uganda, Malawi, Mozambique, Zambia, Zimbabwe, Congo and South Africa.
- ✓ **In the Arabian Peninsula:** Yemen.

Khat is widely cultivated in these areas, as well as in Madagascar. Unlike other drugs, khat has not enjoyed a significant global spread, as only consumption of the fresh leaves containing the active ingredient is effective. This means that consumers need to be located close to production sites to gain access to the plant.

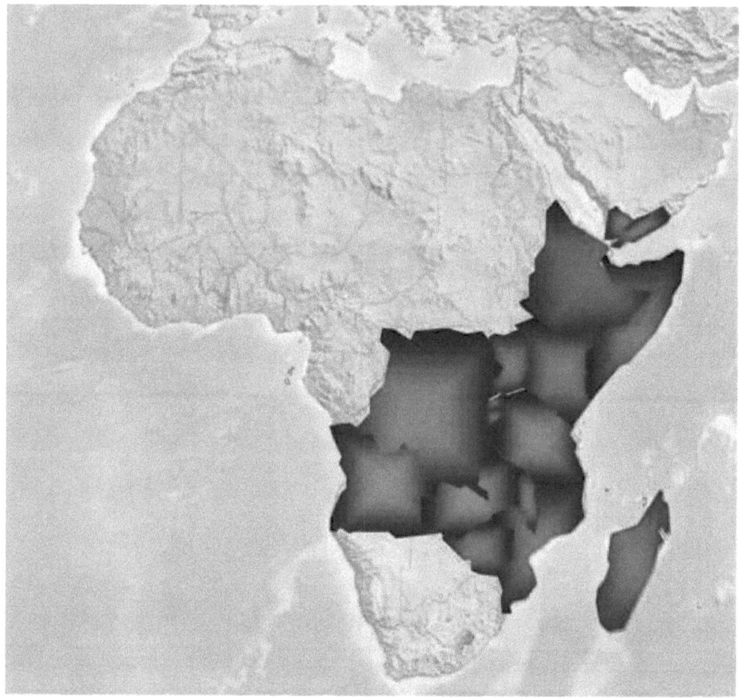

Figure 1: Khat distribution area (consumption and cultivation)

PART II
THE KHAT PLANT

The khat plant

Photo 2: Khat plant (*Catha edulis* (Vahl) Forssk. ex Endl)

1. Khat classification

The *Catha edulis* (Vahl) Forssk. ex Endl plant is classified as follows:
- ✓ **Kingdom:** Plantae
- ✓ **Division:** Magnoliophyta
- ✓ **Class:** Magnoliopsida
- ✓ **Order:** Celastrales
- ✓ **Family:** Celastraceae
- ✓ **Genre:** *Catha*
- ✓ **Species:** *edulis*

APG *(Angiosperm Phylogeny Group)* classification:
- ✓ **Clade:** Angiosperms
- ✓ **Clade:** Dicotyledons
- ✓ **Clade:** Rosaceae
- ✓ **Clade:** Fabidae

- ✓ **Order:** Celastral
- ✓ **Family:** Celastraceae

2. Introducing the Celastraceae family

The Celastraceae family is found mainly in tropical and subtropical regions.

Vegetative system of the khat :

This family includes a variety of trees and shrubs, both deciduous and evergreen. These include climbing species such as *Salacia* and voluble species such as *Hippocratea*. Some species have woody lianas, and aerial roots are sometimes seen in the *Celastrus* genus. Spiny species, such as those in the *Maytenus* genus, are also found, and some trees have buttresses. The root system of Celastracae can reach depths of up to 5 metres.

Celastraceae leaves are generally opposite or pseudo-opposite (with the exception of the *Cassine* genus, where they are alternate). They are petiolate, ex-stipulate or with reduced, deciduous stipules. Leaf blades are simple and often leathery. The venation is pinnate and reticulated, with margins that may be entire, crenate, serrate or toothed.

2.1 Khat reproduction

Khat inflorescences can be axillary or terminal, with a wide variety of cymes, racemes, thyrses, fascicles, or sometimes panicles or solitary flowers. Flowers are actinomorphic and may be bi- or unisexual.

They are generally small and greenish. Flowers have 3 to 5 overlapping sepals, which may be free or fused at the base. The 3 to 5 petals are usually free and inserted on or under a fleshy, glandular nectariferous disk, although this disk is almost absent in *Microtropis*.

The stamens, 3 to 5 in number, are inserted on the nectariferous disk and are free and alternipetal. Anthers are bilocular or sometimes unilocular, basifixed or dorsifixed, with longitudinal or transverse dehiscence in *Hippocratea*, and may be intralescent, extralescent or latrorse.

The ovary is superect or semi-inferior, with 2 to 5 carpels fused into 2 to 5, sometimes incomplete, lodges. Each compartment contains 2 erect or pendulous ovules on axillary placentas. The style is single, terminal and very short, sometimes absent, with a captive or lobed stigma (2 to 5 lobes), corresponding to the number of carpels in the gynoecium.

The fruit may be a schizocarpic capsule with loculicidal dehiscence, releasing 2 to 5 indehiscent mericarps, or more rarely a samara, berry, drupe, or exceptionally, an indehiscent capsule or nut with a lateral style. Some *Euonymus* capsules may have spiny outgrowths. The pericarp may be leathery or fleshy, smooth, angular or with spiny wings. The seeds, 2 to 12 in number, may be smooth or rough, albuminous or exalbuminous, with a large, straight embryo. They often have a colorful basal aril that helps them to be dispersed by birds. Cotyledons are flat and connate. Germination is epigeous.

3. Le Genre *Catha*

Synonyms: *Celastrus edulis* Vahl (1790), *Catha inermis* J. F. Gmel (1791)

Common names: Khat, kat, qat, Arab tea, Abyssinian tea, bushman tea, katyna, mlonge, miraa, murungu.

The *Catha* genus comprises a single species, *Catha edulis*, which is highly polymorphic. There are no recognized sub-specific taxa within this species, although several forms are cultivated. In Ethiopia, for example, farmers distinguish three cultivars:
- ✓ "Dallota" or "Ahde", with small pale green, almost white leaves;
- ✓ "Dimma", with medium-sized red leaves;
- ✓ "Mohedella" or "Hamarcot", with olive-green leaves.

For each of these three cultivars, there are four grades: urata, qudaa, qarxii and faquaa, the latter also known as hiraa or tacharoo. These grades differ in the way the leaves are packed.

Photo 3: Different qualities of khat sold in Ethiopia

In Yemen, khat cultivars are sometimes named after localities, such as "Sabr", "Reimi", "Taizi" and "Mathani". Yemeni cultivar names also sometimes reflect the color of the khat grown. Local cultivars are defined according to their geographical origin, cultivation methods and morphological characteristics, such as leaf color, trunk size and strength of effect. More specifically, four predominant cultivars are easily recognized by farmers in Yemen. The "Abyadh" cultivar, identifiable by its pale green color, is the most widespread.

The other three cultivars are: "Azraq", purplish in color, "Aswad", purple, and "Ahmar", an intermediate form between the last two in terms of color.

Photo 4: The four cultivars present in Yemen

4. Botanical characteristics and cultivation of khat

4.1 Botanical description of khat

Khat is an evergreen, hairless tree that can reach heights of up to 25 meters in its natural habitat. However, when cultivated, it generally takes the form of a multi-stemmed shrub, with a maximum height of 6 meters. The trunk is straight and slender, with a bark that differs according to the growing environment: smooth and pale grey-green in cultivation, it is rough on taller trees in the wild.

Cylindrical branches are grayish to brownish, while young twigs are often flattened and range in color from dull green to brownish red.

Khat leaves are arranged alternately on orthotropic branches and opposite on plagiotropic branches. The stipules, triangular in shape and measuring around 3 cm by 1 mm, are pale green and deciduous, leaving a beaded scar when they fall off. Stalks, cylindrical and 3 to 11 mm long, vary in color from pale to dark green.

16

Leaf blades are generally oblong, elliptical or obovate, with dimensions ranging from 5.5 x 1.5 cm to 11 x 6 cm. The leaf base is wedge-shaped or attenuated, while the apex may be acute, acuminate or even obtuse. Leaf margins are glandular, crenate or serrated, and have a shiny appearance. Mature leaves are leathery with reticulated veins.

Photo 5: Young khat shoots (*Catha edulis*)

Photo 6: Khat leaves (*Catha edulis*)

Khat inflorescences are axillary cymes, often dichasial, up to 3.5 cm long. They bear numerous flowers, and the stalk can be up to 12 mm long. The bracts are generally triangular and persistent.

Flowers are bisexual, regular and pentamerous, with a diameter ranging from 2 to 4 mm. The pedicel is 1 to 2.5 mm long. The sepals, connate at the base, are broadly oval or suborbicular with fimbriate margins.

The petals are free, elliptical and oblong, with finely crimped or fimbriated edges and colors ranging from white to pale yellow. The stamens are also free, alternating with the petals and slightly shorter than the latter. The intrastaminal disk is fleshy and slightly 5-lobed.

The ovary, upper and broadly ovoid, is trilocular and has three short styles with small stigmas.

Photo 7: Inflorescences of khat (*Catha edulis*)

The khat fruit is a trigonal, drooping, narrowly oblong capsule, 6 to 12 mm long. Its color varies from red to brown. It is a dehiscent, loculicidal fruit with three valves, containing one to three seeds.

Figure 2: Botanical plates of khat (*Catha edulis*)

(Legend: 1: sterile orthotropic branch; 2, A: flowering plagiotropic branch; 3, B: flower ;

C: cupped flower; D: cupped fruit; 4, F: dehiscent fruit; 5, G: seed ;

E: Seed-bearing branch)

Photo 8: Khat branches with fruit

4.2 Khat ecology

20

Khat is grown mainly in Yemen, Ethiopia and Kenya, where climatic and environmental conditions are particularly favorable to its development. Its natural habitat lies between latitudes 18°N and 30°S. The plant thrives on high plateaus at altitudes of between 1,500 and 2,000 meters, where average daily temperatures range from 16 to 22°C. These conditions are ideal for growing khat.

Annual rainfall requirements for khat vary between 800 and 1,000 mm, spread over a period of 4 to 6 months. By comparison, sunflowers require around 420 mm of water per year. Frost and excessive humidity are limiting factors for khat cultivation. Khat prefers moderately acidic to slightly alkaline soils. However, it can adapt to different types of soil, from sandy to clayey, provided they are sufficiently deep and well drained. A high level of organic matter in the surface layers of the soil is also essential for optimum plant growth.

Photo 9: Photo of khat (*Catha edulis*) showing opposing branches

Photo 10: Cultivation of khat (*Catha edulis*) on terraces

4.3 Khat plantation

Khat is propagated mainly by cuttings. Cuttings, measuring between 30 and 50 cm, are taken from orthotropic branches or suckers close to the ground. They root easily, either directly in the field or in nurseries during the rainy season.

Seedlings are generally planted in rows, either on gentle slopes or on flat land, hence the frequent use of terraces for cultivation. In the event of insufficient rainfall, an irrigation system is required to ensure optimum growth. Mixed cropping is also common: some plots of khat are alternated with rows of coffee (*Coffea arabica* L.), optimizing the use of resources and diversifying crops.

Photo 11: Mixed cultivation of khat and coffee trees

4.4 Khat crop management

Khat generally grows without major intervention for the first 3 to 4 years, reaching a height of around one meter. Normal yields are reached between 5 and 8 years after planting. Plant height is regulated by regular pruning, generally between 2.5 and 5 meters. With proper care, plantations can maintain productivity for up to 75 years. In times of drought, irrigation can be introduced to enable out-of-season harvesting, which can be economically advantageous due to higher market prices. As far as disease is concerned, khat is relatively resistant, with infrequent and mild diseases.

4.5 Khat harvest

Khat is harvested exclusively in the early morning to preserve the freshness of the leaves. This operation is carried out 2 or 3 times a week during the harvesting season. The young shoots are cut to a length of around 40 cm and then grouped into bunches of appropriate size for a two-hour chewing session, representing around 500 g per bunch. The harvested leaves represent around 150 g per bundle.

Yields from crops destined for Yemeni markets can reach up to 2 tonnes per hectare per year.

4.6 Post-harvest treatment of khat

After harvesting, the bunches of khat are first sprayed with water to maintain their freshness. They are then wrapped in banana leaves and packed in plastic bags for transport to market. This method of processing preserves the freshness of the young shoots, essential as khat is a rapidly perishable commodity. Indeed, after 24 hours, the bundles lose their stimulating effect, which can reduce their value on the market.

The quality of khat can vary considerably, and merchants and consumers alike evaluate it according to several criteria: the origin of the plant, the time of harvest, and the color and tenderness of the leaves. Color is particularly important in judging product quality. In general, whitish leaves are considered to be of superior quality. However, a reddish tinge to the leaves is often associated with a stronger stimulant effect.

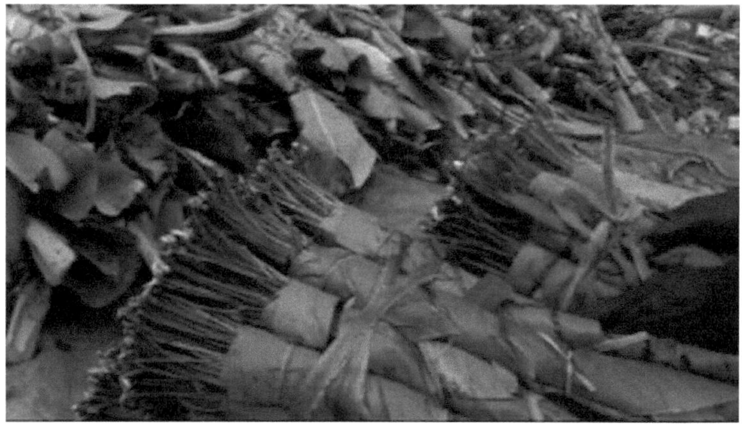

Photo 12: Khat bundles wrapped in a banana leaf

PART III

THE BIOCHEMISTRY OF KHAT

1. Khat composition

Fresh khat leaves contain a significant diversity of chemical compounds, including elements typical of plants. There are over forty alkaloids, glycosides, amino acids, vitamins and minerals. Leaf composition comprises 90% water, 1.6% tannins (polyphenols), 5-6% protein, 2-3% fiber, 0.3% calcium, 0.2% vitamin C, as well as terpenes, flavonoids and sterols. Among these compounds, two alkaloids are particularly noteworthy. Khat's main active ingredient is S-cathinone, also known as (-)-2-aminopropiophenone or S-(-)-2-amino-1-phenyl-1-propanone, a β-keto derivative of amphetamine.

Various methods are used to separate and analyze the alkaloids present in khat. Thin-layer chromatography, followed by gas chromatography-mass spectrometry, is particularly effective. However, the use of high-performance liquid chromatography (HPLC) has also been reported.

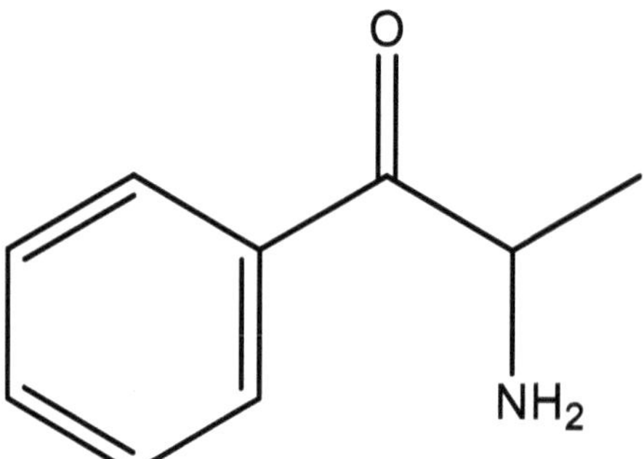

Figure 3: Molecular structure of cathinone

Cathinone is a highly unstable substance. In fact, between 24 and 36 hours after harvesting, it transforms into a dimer, whose activity is much lower than that of the original cathinone. This is why khat is eaten fresh to preserve its stimulating effects.

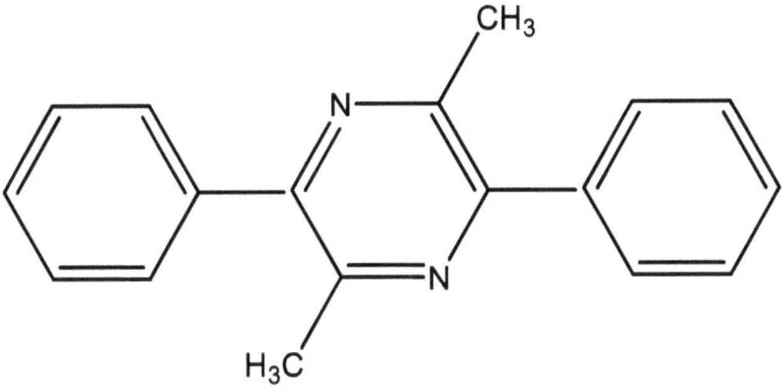

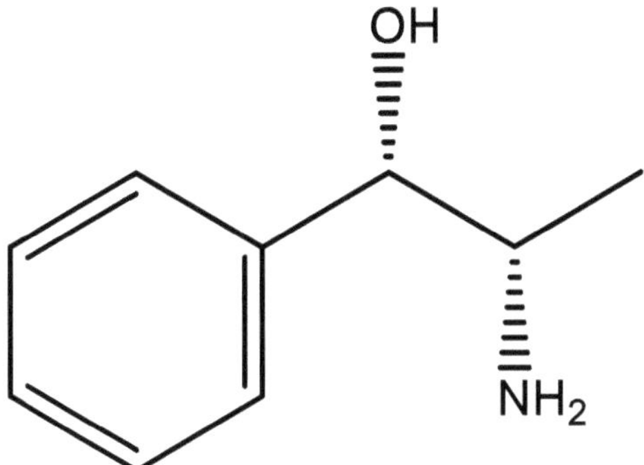

Figure 4: Molecular structure of the cathinone dimer

The second alkaloid contributing to khat's desired effects is cathinone, also known as 1S, 2S-norpseudoephedrine. This psychoactive substance is derived from the metabolism of cathinone within the plant. Cathinone is metabolized into cathine and norephedrine. Although cathine plays a role in khat's effects, it is ten times less active than cathinone.

Figure 5: Molecular structure of cathine

The khat plant also contains trace amounts of other substances. These include 1R, 2S-norephedrine, as well as a large number of cathedulins, polyhydroxylated sesquiterpenes, and several phenylalkylamines such as phenylpentenylamine, merucathinone, pseudomerucathinone and merucathine, which have no significant impact on the plant's stimulant activity. Quinone triterpenes are also present in the roots, giving them an orange-red color.

The cathinone content of fresh khat leaves is around 0.1%. Quantities of cathine are lower in fresh plants, as cathinone is converted to cathine during ripening. Khat leaves contain around four times more cathine than norephedrine. Concentrations of cathinone, cathine and norephedrine vary considerably between khat species. For 100 g of fresh leaves, cathinone can vary from 36 to 383 mg, cathine from 83 to 120 mg, and norephedrine from 8 to 47 mg. These wide variations can also be influenced by the degree of freshness of the samples.

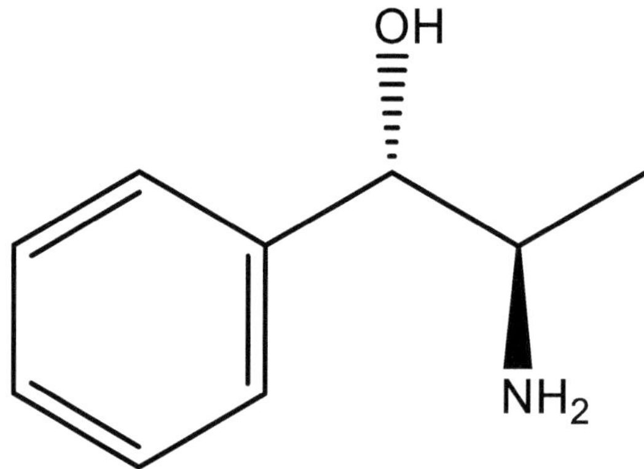

Figure 6: Molecular structure of norephedrine

Norephedrine is the third main alkaloid in khat. Like cathine, it results from the degradation of cathinone. It is in fact a diastereomer of cathine. Norephedrine contributes to some of the effects of khat, including adverse effects such as headaches.

2. Pharmacological activity of Khat

The alkaloids in khat, cathinone and cathine, are central nervous system stimulants with amphetamine-like effects.

Cathinone and cathine act primarily on the dopaminergic and noradrenergic pathways. Like amphetamine, cathinone stimulates serotonin secretion in the central nervous system and also increases dopamine release, thereby enhancing dopaminergic activity. In addition, cathinone influences noradrenaline storage sites, facilitating noradrenergic transmission. Cathinone and cathine also appear to inhibit noradrenaline reuptake.

Despite these stimulating effects, the potency of cathinone and cathine is significantly lower than that of amphetamines. Cathinone is about half as potent as amphetamine, while cathine is seven to ten times less potent. Consequently, the amount of khat needed to achieve a mild stimulant effect is considerably greater, making khat less dangerous than pure psychoactive drugs.

Khat consumption is generally associated with a low degree of dependence. The desired effects of khat alkaloids include euphoria, exhilaration, lucidity and heightened excitement. However, these stimulating effects can be followed by depression, irritability, anorexia and sleep disorders. Psychotic reactions have been observed in frequent users at high doses. Symptoms such as hyperthermia and analgesia suggest activation of monoaminergic pathways and opioid mechanisms, while hyperthermia may also result from stimulation of the thyroid gland.

The role of other khat components is less well understood. Nevertheless, it has been suggested that astringent tannins may be responsible for gastritis, stomatitis, esophagitis and periodontitis. On the other hand, the high vitamin C content of fresh leaves may confer some nutritional value.

3. Khat pharmacokinetics

The stimulant effects of khat appear about one hour after the start of chewing. Plasma concentrations of alkaloids peak around 1.5 to 3.5 hours after the start of chewing. After chewing 60 g of fresh plant for one hour, the average plasma concentration of cathinone can reach up to 100 ng/ml. Cathinone is almost completely eliminated from the blood eight hours after chewing, due to a first-pass hepatic effect that leads to the formation of norephedrine, excreted mainly via the urinary tract. Only 2-7% of cathinone is excreted unchanged. Cathinone is mainly excreted unchanged.

A two-compartment pharmacokinetic model has recently been described. It comprises two phases of absorption: the first lasting 0.1 to 0.2 hours, the second 1 to 2 hours. Absorption through the oral mucosa represents the first absorption segment, during which the majority of cathinone (59 ± 21%) and cathine (84 ± 6%) is absorbed. Peak plasma levels are reached in 2.3 hours for cathinone, 2.6 hours for cathine, and 2.8 hours for norephedrine. Chewing extracts a large proportion of khat's active ingredients, leaving only 9.1 ± 4.2% in leaf residues. The second absorption segment involves the stomach and small intestine.

The elimination half-life of cathinone is 1.5 ± 0.8 hours, while that of cathine is 5.2 ± 3.4 hours. Central compartment volume is 2.7 ± 3.4 l/kg for cathinone and 0.7 ± 0.4 l/kg for cathine. Cathinone is detectable in plasma between 0.5 and 7.5 hours after administration, but generally disappears after 24 hours.

Gas chromatography/mass spectrometry is used to detect the presence of khat in urine. As cathinone cannot be detected directly, we look for its metabolite, norephedrine. The test can remain positive for up to two days after khat consumption.

In addition, khat can influence the bioavailability of certain drugs, such as ampicillin, amoxicillin and tetracyclines. A study carried out at Sana'a University showed a significant reduction in the bioavailability of ampicillin when taken at the same time as khat.

However, ampicillin administered two hours after khat consumption is not affected. Amoxicillin bioavailability is significantly reduced only when administered during the khat session. It is therefore recommended to take these antibiotics two hours after a khat session. This study focuses on these two antibiotics because of their frequent use in Yemen.

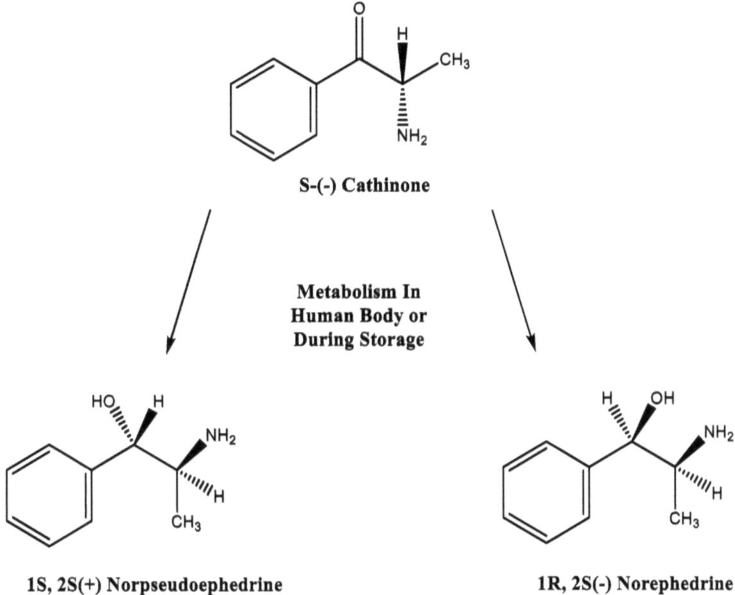

Figure 7: Metabolism of cathinone in the human body

PART IV
EXTENT AND DANGERS OF KHAT

1. Prevalence of Khat

It is estimated that over 80% of men in Yemen regularly consume khat. Female consumption is also on the rise, with around 43-50% of women chewing khat daily in the country.

In Djibouti, khat is consumed by around 70% of men and 30% of women.

In Somalia, consumption is predominantly male, with 75% of men being regular users, while 7-10% of women also use khat. Some 15-20% of children under the age of 12 are also daily users.

In Ethiopia, khat is consumed by around 50% of the population, 17% of whom claim to be daily consumers.

In Europe, khat consumption is relatively sporadic, mainly affecting immigrant communities from Djibouti, Somalia, Ethiopia, Kenya and Yemen. Available studies, mainly carried out in the UK, do not allow us to determine the precise extent of this consumption. However, it seems that consumption is regular but not necessarily daily, with a higher prevalence among men who group together for chewing sessions. Female consumption may be underestimated due to the stigma associated with the practice, leading to consumption in private spaces. Moreover, khat consumption is more widespread among Muslims, as the habit is often passed down from generation to generation. There is also a correlation between student status and khat consumption, particularly during exam periods, due to its stimulating effects on attention.

Contrary to trends in Europe, where poly-drug use is increasingly common, khat users do not generally tend to use other drugs. However, khat is also consumed in several Western countries such as Australia, the USA, Canada, the UK, Denmark, Norway, Sweden and the Netherlands. For example, there are around 1,350 users in Denmark and 2,000 to 3,000 in Sweden.

In these countries, consumption is mainly confined to immigrant communities from Somalia, Ethiopia, Yemen and East Africa in general. Overall, the number of khat consumers worldwide is estimated at around 10 million.

The following diagram (Figure 8) shows how khat reaches Europe.

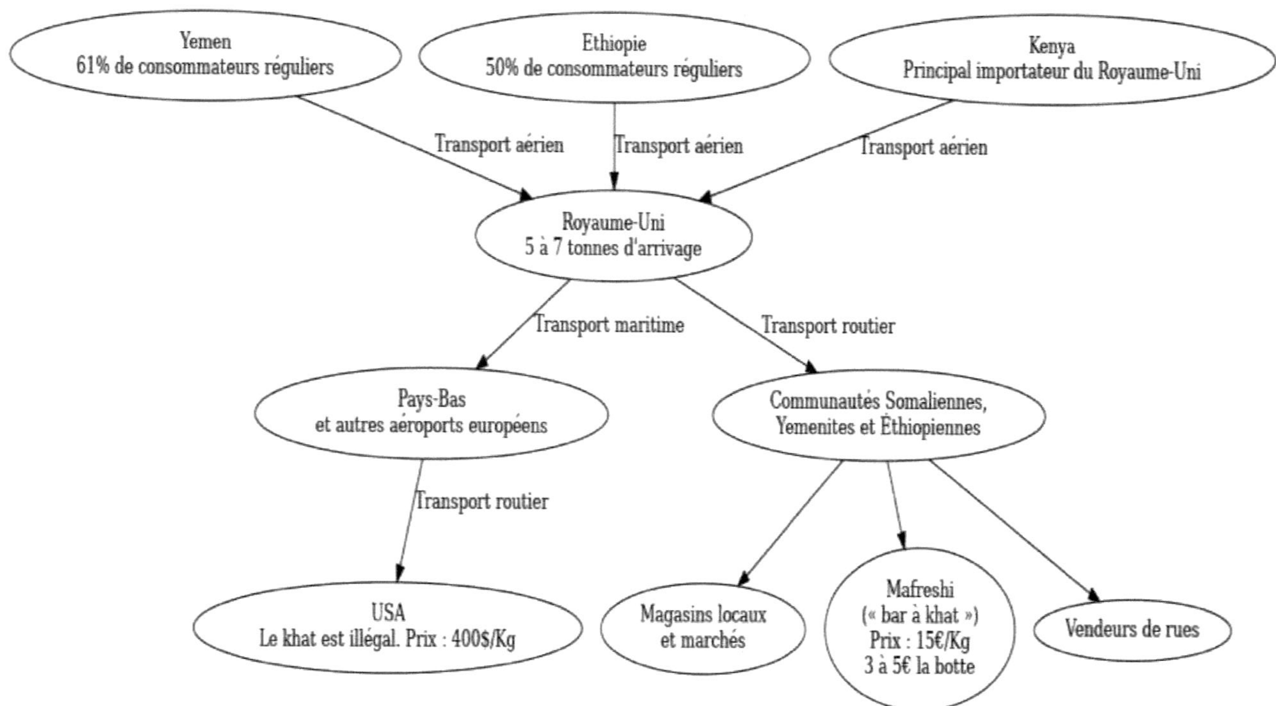

Yemen
61% de consommateurs réguliers
Ethiopie
50% de consommateurs réguliers
Kenya
Principal importateur du Royaume-Uni
Transport aérien
Transport aérien
Transport aérien
Royaume-Uni
5 à 7 tonnes d'arrivage
Transport maritime
Transport routier
Pays-Bas
et autres aéroports européens
Communautés Somaliennes, Yemenites et Éthiopiennes
Transport routier
USA
Le khat est illégal. Prix : 400$/Kg
Magasins locaux et marchés
Mafreshi
(« bar à khat »)
Prix : 15€/Kg
3 à 5€ la botte
Vendeurs de rues

2. Effects and consequences of Khat

Khat consumption is mainly associated with euphoric effects, but it also has short-, medium- and long-term side effects. In the short term, users may experience euphoria, increased stimulation and temporary mood enhancement. With prolonged use, however, a variety of side effects can occur, including loss of motivation, lethargy, depression and mild tremors. Although these symptoms are generally mild, they are often reversible on discontinuation.

The abuse and dependence potential of khat is considered relatively low. However, a number of studies have highlighted various deleterious effects of khat on the body, which we will examine in greater detail.

2.1 Effects of Khat on the Central Nervous System

Consumption of khat, particularly due to the presence of cathinone, results in significant stimulation of the central nervous system. Cathinone, being fat-soluble, reaches the central nervous system more rapidly and intensely than cathine. The psychostimulant effects of khat chewing include moderate euphoria, mild excitement, increased alertness, and improved social interaction and loquacity. These effects generally begin between 15 and 45 minutes after the start of chewing and can last up to 7 hours, with peak intensity observed 1.5 to 3.5 hours after the start of the session. However, these effects are followed by symptoms such as dysphoria, anxiety, reactive depression, insomnia and anorexia.

The effects of khat, although similar to those of amphetamine, differ quantitatively. Khat consumption can induce behaviors such as hyperactivity and logorrhea. A study published in Frontiers in Psychology compared the ability of khat users and non-users to respond to a model signal. This study revealed that, although there were no significant differences between the two groups, khat consumers took longer to respond.

In Europe, khat-related psychoses have become more frequent in recent years. These psychoses can occur following excessive consumption or more concentrated plants, especially in individuals with psychiatric predispositions. Psychiatric disorders observed include mania (such as obsessions), schizophrenia, and paranoid delusions. Hallucinations are not uncommon among regular users. Nevertheless, moderate consumption does not appear to significantly affect psychiatric morbidity.

Moral suffering is also associated with khat consumption, often linked to the frequency and duration of consumption. This relationship is complex, as many people use khat to relieve this suffering. Consumers of more than two bundles of khat a day show increased psychiatric morbidity, and adverse effects are generally dose-dependent. Khat is sometimes considered a trigger for psychiatric illness, with violent episodes reported in the context of paranoia and persecutory delusions.

With regard to dependence, a group of experts from the World Health Organization (WHO) concluded that psychic dependence on khat is significantly greater than physical dependence. After a prolonged session, symptoms include lethargy, moderate depression and recurrent nightmares. The absence of physical symptoms suggests that the effects of stopping consumption are more of a rebound type than a true withdrawal syndrome. Nevertheless, khat remains under scrutiny due to the frequent consumption of cigarettes and alcohol during chewing sessions, with alcohol offsetting khat's effects on mood and sleep.

A recent study in Kenya revealed that two-thirds of khat users drink alcohol, 41% of whom are heavy users. What's more, khat-induced insomnia can lead to more dangerous behaviors, such as glue-sniffing or sedative use.

2.2 Cardiovascular effects of Khat

Regular consumption of khat is associated with a significant rise in diastolic blood pressure. In volunteers, an increase in blood pressure correlates with a peak in plasma khat levels, generally reached between 1.5 and 3.5 hours after ingestion. Cathinone, one of the main components of khat, contributes to this rise in blood pressure by exerting a positive inotropic and chronotropic effect on the heart. Studies on anesthetized rats and dogs show that cathinone also increases heart rate.

Khat consumption is also associated with vasoconstriction, amplified by increased noradrenaline release. At peripheral noradrenaline storage sites, cathinone promotes noradrenaline release into the synaptic cleft in a similar way to amphetamine. However, this release is inhibited by noradrenaline reuptake blockers such as cocaine or desipramine. These observations suggest that cathinone acts indirectly as a sympathomimetic.

Reports such as Fitzgerald's also mention cases of pulmonary edema linked to khat use.

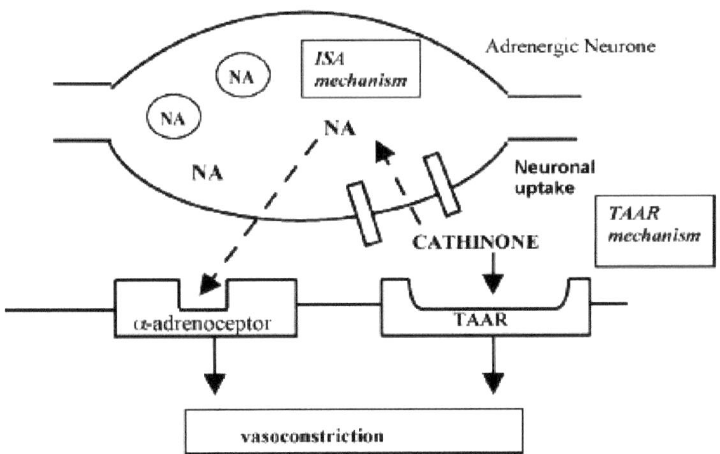

Figure 9: Vasoconstriction mechanism of cathinone

Cathinone may act as an indirect sympathomimetic amine (ISA mechanism), via a reuptake mechanism in the sympathetic neuron; this would cause a noradrenaline release on adrenoreceptors. The second possibility may be via an independent sympathomimetic mechanism, probably directly on an amine-type trace receptor (TAAR mechanism).

Individuals taking khat experience an increase in heart rate and body temperature, as well as profuse sweating. These effects are often accompanied by coldness of the extremities, a clinical sign of peripheral vasoconstriction.

2.2.1 Myocardial infarction and Khat

In recent years, khat consumption habits have evolved, with chewing sessions often extending well into the evening, sometimes until midnight. This change influences the circadian rhythm of cardiovascular events. In general, myocardial infarctions and sudden deaths occur more frequently early in the morning, just after waking up. This is linked to an increase in catecholamines, which raise heart rate, blood pressure, myocardial contractility and oxygen demand first thing in the morning.

Khat consumption is an independent risk factor for cardiovascular events. People who chew khat moderately have an increased risk of heart attack, while those who chew khat intensively have an even higher risk. The duration of chewing sessions is directly correlated with heart attack risk: individuals who chew khat for more than six hours are at increased risk. In addition, cigarette smoking among khat users can complicate this assessment, as it is frequent and can also influence the risk of cardiovascular disease.

Cathinone, the main active compound in khat, causes vasoconstriction of the coronary arteries. This vasoconstriction could explain the increase in cardiovascular accidents observed among khat users.
In the case of acute coronary syndrome, khat consumption is directly associated with an increased risk of heart attack and death; for example, a significant proportion of acute coronary syndrome patients in Yemen are khat consumers. The increased heart rate, coronary spasm and exacerbation associated with khat consumption also contribute to the risk of heart attack.

It is crucial to study these risks in different populations, comparing non-smoking khat consumers, smoking khat consumers, non-smoking non-consumers and smoking non-consumers.

2.2.2 Khat-related strokes

Some studies suggest a link between khat consumption and strokes. The cathinone-induced vasoconstriction associated with khat-induced hypertension, as well as the myocardial infarctions observed in users, may indicate that similar mechanisms are at work in the brain.

Cases of stroke following khat consumption have been documented. For example, a 41-year-old man suffered cerebral ischemia after consuming khat. In another case, a 28-year-old man suffered both a myocardial infarction and a stroke after a khat chewing session. These incidents were attributed to a thrombogenic mechanism, where blood clot formation could play a crucial role.

Khat consumption has been shown to increase the risk of stroke. However, further research is needed to confirm and better understand the link between khat consumption and stroke.

2.2.3 Thrombosis, Aspirin intake and Khat

Aspirin is commonly used to prevent cardiovascular events, both in healthy people and in patients with ischemic heart disease. A low dose of aspirin, usually 75 mg, is prescribed to reduce the risk of heart attacks and strokes.

However, studies have shown that in individuals consuming khat, bleeding time is significantly reduced compared with non-consumers. Indeed, patients taking aspirin and consuming khat had a bleeding time of around 2.3 minutes, compared with 8 minutes for those not consuming khat. These results suggest that the components of khat may attenuate the antiplatelet effects of aspirin.

It is possible that cathinone, one of khat's active ingredients, exerts a pro-aggregating effect by stimulating the release of catecholamines. However, these hypotheses still need to be confirmed by *in vitro* studies and clinical research to better understand the interaction between khat and anticoagulant drugs such as aspirin.

2.3 Effects of Khat on the Digestive System

The effects of khat on the digestive system can be seen in the various pathologies observed in consumers. Among the disorders frequently reported are stomatitis, esophagitis and gastritis. These symptoms may be due to the tannins present in khat, known for their astringent properties. Gastric disorders are also associated with gastric hypotonia, resulting from the sympathomimetic action of cathins and their precursors. Recent studies have shown that chewing khat significantly slows gastric emptying, contributing to an increase in gastro-oesophageal reflux. Acid regurgitation and retrosternal pain are therefore common clinical manifestations, and acid reflux increases the risk of Barrett's Esophagus.

Research has also linked khat consumption to an increased risk of duodenal ulcer, taking into account other risk factors such as smoking, non-steroidal anti-inflammatory drugs (NSAIDs) and alcohol. Although these factors have been ruled out as causes of ulcers, the presence of *Helicobacter pylori* and other variables, such as pesticides or chemicals on khat leaves, as well as khat constituents such as cathinone, have been suggested as potential causes.
One study also showed that high doses of khat (300 mg/kg) aggravated existing gastric and duodenal ulcers, attributing this harmful activity to the dose, the plant's chemical constituents, and its amphetamine-like effects. At lower doses or with sub-chronic administration, no significant worsening of ulcers is observed.

Khat consumption is also associated with anorexic effects. Users tend to have inadequate meals on the day of the session, the anorectic effects being attributed to a direct action of cathinone on the central nervous system and gastrointestinal tract. Prolonged use at high doses results in significant weight loss and anorectic activity.

In addition, frequent consumer complaints include constipation, due to the astringent effect of tannins and the sympathomimetic properties of cathinone. This constipation is generally dose-dependent, and more pronounced with prolonged use of high doses. Users often try to compensate for these effects by modifying their diet, for example by eating high-fat meals before sessions.

Chronic khat consumption is also associated with an increase in hemorrhoidal attacks. One study found that 62% of consumers suffered from hemorrhoidal attacks, compared with just 4% of non-consumers. However, this study does not take into account other potential factors, such as smoking, diet, age, or other disorders such as constipation. The origin of hemorrhoids could be due to the cathinone or tannins present in khat.

The effects of khat on the liver are also a cause for concern. The liver appears to be particularly vulnerable to the toxic effects of khat, affecting levels of alkaline phosphatase and alanine aminotransferase (ALAT). Consumption of khat leads to biochemical and histological alterations, both in the short and long term, such as congestion of the hepatic central vein, hepatocellular degeneration, and regenerative activity leading to fibrosis, which can progress to cirrhosis. Cathinone-induced vasoconstriction could also contribute to certain liver pathologies, such as subchronic hepatitis and autoimmune hepatitis.
Jaundice, typical of drug-induced hepatitis, has also been observed. Hepatotoxicity may result from increased oxidation, increasing lipid peroxidation and free radical production in the liver. However, other studies have not revealed significant changes.

2.4 Khat, Type II Diabetes and Effect on Appetite

Consumption of khat, in particular because of the cathinone, has notable effects on blood glucose regulation and appetite. Cathinone stimulates the secretion of noradrenaline and adrenaline, which bind to $\beta2$-adrenergic receptors in the liver, activating glycogenolysis and raising blood glucose levels. At the same time, these catecholamines bind to $\alpha2$-adrenergic receptors in the pancreas, inhibiting insulin secretion and increasing muscle glucose utilization. This dual action contributes to a rise in blood glucose levels.

In Yemen, many diabetics believe that eating khat helps control their blood sugar levels. Moreover, khat is often seen as a means of managing hyperglycemia. The idea that khat could be beneficial for Type I diabetics is based on its weight-loss effect, linked to lipolysis and delayed gastric emptying. Sympathetic stimulation by cathinone leads to increased plasma levels of catecholamines, which can raise blood glucose levels by activating glycogenolysis in muscle and liver. Consequently, inhibition of insulin secretion by pancreatic β-cells may contribute to elevated blood glucose levels.

However, study results are contradictory. Some studies have shown no significant effect of khat on immediate and post-prandial glucose levels in non-diabetic subjects. Conversely, other studies indicate that khat consumption may reduce blood glucose levels. For example, one study observed a 61% reduction in blood glucose four hours after a chewing session in healthy subjects. In diabetics, on the other hand, blood sugar levels were significantly higher one to two hours after chewing.
Research on rabbits fed different doses of khat showed an increase in blood glucose levels after four months, followed by a significant reduction after six months. These results are complex to interpret, but suggest a relationship between the stimulation of glycogenolysis and the effect on insulin secretion, with a possible compensation of the rise in blood sugar by an increase in insulin secretion. In addition, studies on rabbits revealed that khat could lower cholesterol levels after six months, which could reduce cardiovascular risk.

A recent study has established a clear link between regular khat consumption and insulin-requiring type II diabetes. Markers considered include resistin, cortisol, insulin, as well as blood copper, zinc and calcium levels. The results show that khat consumption increases insulin resistance in type II diabetics; this is associated with elevated immediate and post-prandial blood glucose levels, as well as changes in cortisol, copper, calcium, zinc and insulin levels.

With regard to appetite, khat exerts an anorectic effect by significantly reducing the sensation of hunger and increasing satiety, although ghrelin and peptide YY levels are not affected. This anorectic effect is attributed to a central action of cathinone. Thus, khat could be used to control or reduce obesity and indirectly the risk of diabetes in certain populations. In addition, an increase in levels of the anorectic hormone leptin was observed four hours after an intense chewing session (400 g of khat), helping to reduce appetite and, consequently, weight.

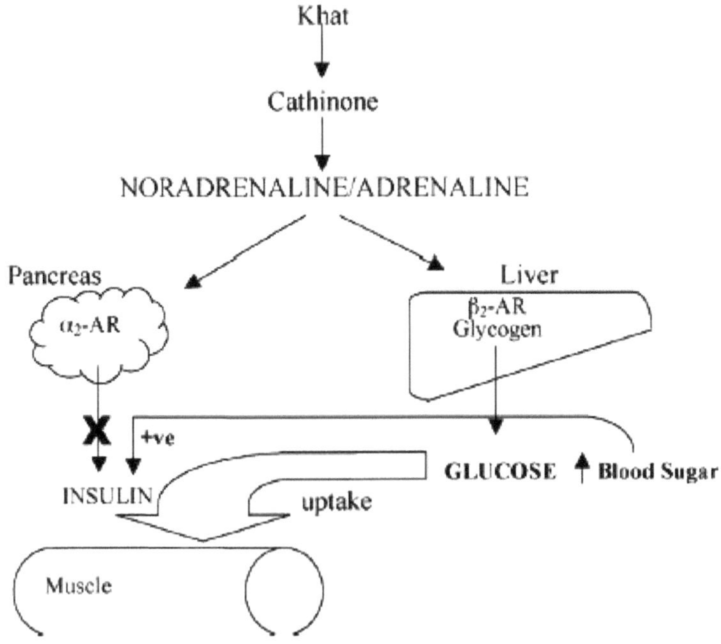

Figure 10: Schematic diagram of the hypothetical effect of khat and cathinone on blood glucose levels

2.5 Khat and the genitourinary system

2.5.1 Effects of Khat on humans

Khat consumption in men is associated with various effects on the genitourinary system, including urination problems and impaired renal and reproductive function.

Micturition and kidney function: Khat users may experience difficulties with micturition, such as hesitant urination and low flow. These symptoms are probably due to the stimulation of $\alpha 1$-adrenergic receptors in the bladder by cathinone, a sympathomimetic alkaloid.

This urination problem can be alleviated by indoramine, a drug which does not, however, affect the cardiac effects of cathinone. Furthermore, khat consumption is associated with elevated urea and creatinine levels, indicating renal toxicity. The accumulation of free radicals in kidney tissue, generated by lipid peroxidation and other oxidation mechanisms, also contributes to this toxicity.

Sexual and reproductive function: The effects of khat on libido and reproductive function are varied and sometimes contradictory. Some users report improved self-esteem and libido. However, studies on the effects of khat on sexual function and sperm production do not allow definitive conclusions to be drawn.

Research indicates that cathine and norephedrine, present in khat, may influence spermatogenesis. Initially, these compounds appear to accelerate the maturation of spermatids into spermatozoa by facilitating acrosome formation. In addition, they may help maintain the acrosome, thereby promoting fertilization.

However, other studies, such as that by Mwenda *et al.* show that khat consumption can impair spermatogenesis and reduce plasma testosterone concentrations. Moreover, according to Mohammed et *al.* a moderate dose of khat (100-200 mg/kg) may increase sexual motivation without significantly affecting performance, while higher doses (300 mg/kg) decrease both motivation and sexual performance.

In chronic khat chewers, several studies have observed a reduction in sperm count, volume and motility. Sperm malformations have also been documented in Yemen among regular consumers, such as sperm without flagella, with a headless flagella, or with multiple heads and flagella.

2.5.2 Effects of Khat on women

Research into the effects of khat on women, particularly during pregnancy and breastfeeding, reveals several negative impacts on maternal and neonatal health.

Impact on Pregnancy and the Newborn: Studies show that children born to mothers who consume khat, whether occasionally or regularly, often have lower-than-average birth weights. Recently, khat consumption during pregnancy has been shown to have deleterious effects on several neonatal parameters. These include a reduction in weight, height and head circumference, as well as a decrease in the Apgar index, which assesses a newborn's state of health based on heart rate, breathing, skin coloration, muscle tone and response to stimulation.

These effects are proportional to the intensity of the mother's khat consumption. Research indicates that frequent khat consumption impairs intra-uterine fetal growth by affecting utero-placental blood flow. For example, a study by Hassan et *al.* on rats showed that khat reduced fetal body fat and weight, while altering the chemical composition of certain fetal organs, such as the liver, heart and kidneys. These effects are attributed to a suppression of protein synthesis in these organs.

Moreover, khat has genotoxic and teratogenic properties, particularly in regular pregnant users. Studies on mice and rats have revealed lethal mutations, chromosomal aberrations in sperm cells and teratogenic effects.

Impact on breast-feeding: It has been observed that breast-feeding women who consume khat generally produce a reduced quantity of milk. This reduction in lactation may be due to inhibition of prolactin secretion by cathine, an active component of khat. Moreover, cathine is detected in the breast milk of consumers. Traces of this substance can also be found in infants' urine 2 to 4 hours after breast-feeding.

2.6 Khat and Cancer

2.6.1 Cancer localization

Prolonged consumption of khat can have chronic deleterious effects, particularly on cancers. The way this plant is consumed has a strong influence on the location of the cancers observed, which are mainly associated with the oral cavity and digestive tract, with a clear dependence on the dose consumed.

✓ **Oral cavity :**

Cancers of the oral cavity, such as those affecting the lower jaw, oral mucosa and lateral surface of the tongue, generally occur in people who have consumed khat for at least 20 years. Prolonged accumulation of khat is correlated with an increase in cases of oral malignancy, as revealed by epidemiological studies in Saudi Arabia. The tannins present in khat can thicken the oropharyngeal and esophageal mucosa, which may increase the risk of cancer.

✓ **Digestive tract :**

Esophageal and gastric carcinomas were observed in some patients who consumed khat, with women being more affected by esophageal carcinomas. The use of water pipes during khat sessions has been identified as an additional risk factor for cardiac carcinoma. Khat consumption is also associated with delayed gastric emptying, which can lead to gastro-oesophageal reflux, a potential precursor of Barrett's Esophagus and associated cancers.

✓ **Duodenal ulcers :**

Although duodenal ulcers are common among khat users, they also appear to be linked to tobacco consumption. The transformation of esophageal mucosa into gastric-type mucosa is suspected of significantly increasing the risk of esophageal adenocarcinoma. However, this risk is high regardless of khat consumption, and warrants further study to determine the specific factors.

At present, there is insufficient evidence to suggest that khat is the sole carcinogenic factor. Although studies have established a link between khat consumption and cancers of the oral mucosa, the concomitant presence of tobacco and other carcinogenic substances, such as polycyclic aromatic hydrocarbons and nitrosamines, complicates the identification of khat's exclusive role in these conditions. The complexity of the situation lies in the difficulty of distinguishing the carcinogenic effects of khat from those of other environmental and behavioural factors.

2.6.2 Khat, Oxidative Stress and Genotoxic Effects

Research has highlighted the harmful effects of khat on liver and kidney health, notably through studies on animal models. Investigations revealed severe khat toxicity, reflected in increased liver enzyme levels, as well as increased serum concentrations of urea, bilirubin and phosphorus. At the same time, a decrease in total protein and albumin concentrations was observed. This phenomenon is associated with lipid peroxidation and oxidative stress in liver and kidney tissue. However, the administration of antioxidants such as vitamin E or alpha-lipoic acid has shown protective effects against this hepatic toxicity. The toxicity observed seems to be linked to lipid oxidation and free radical formation in these tissues.

With regard to genotoxic effects, studies have shown that methanolic extracts of khat induce chromosomal abnormalities in mouse bone marrow cells. These abnormalities include sister chromatid exchanges, various chromosomal aberrations, as well as disturbances during the metaphase and anaphase phases of cell division. These results underline the genotoxic potential of khat, highlighting the risks of genetic disruption associated with its consumption.

2.6.3 Khat and its anticancer potential

Recent research has explored the anti-cancer potential of khat, revealing promising results. Studies have shown that khat extracts, together with its active components cathinone and cathine, can induce apoptosis in various human cell lines. This induction of apoptosis occurs rapidly and sensitively, thanks to the activation of caspases -1, -3 and -8, which are crucial enzymes in the process of programmed cell death.

Moreover, the effects of khat on cancer cells appear to be comparable to those of camptothecin, a compound known for its anti-cancer properties. These observations suggest that khat could represent an interesting avenue for future research in the field of oncology, particularly for the development of new therapeutic strategies against cancer.

2.7 Khat and oral and dental disorders

In addition to the oral cancer risks associated with khat consumption, the practice is also linked to a number of other oral and dental ailments. Prolonged consumption of khat can lead to stomatitis, often complicated by superinfections of the oral mucosa, due to the mechanical tension exerted on teeth and oral tissues during chewing. The substances contained in khat also have an irritating effect on these mucous membranes. Among khat chewers in Yemen, there is a high rate of gum disease, but a low rate of dental caries. One of the most common complaints is dry mouth, probably caused by the sympathomimetic action of cathinone or by excess salivary secretion linked to chewing.

Studies of Yemeni patients who have consumed khat for around 20 years show a low prevalence of dental caries, but significant tooth wear. Temporomandibular pain and increased periodontal disease on the chewing side were also observed. Around 50% of the cases studied showed oral keratosis, or leukoplakia, a condition considered precancerous.

Khat chewing is also associated with histopathological changes such as mucosal hyperkeratosis (acanthosis). However, some research, such as that carried out in Kenya, does not confirm the association between khat and leukoplakia.

A recent study carried out in Yemen indicates that khat consumption increases the risk of various oral ailments, including gingivitis, gingival pocket formation, gum retraction, tooth loosening and devitalization. Chewing khat also causes temporomandibular cracking and pain, as well as tooth wear and staining.

As far as the salivary glands are concerned, xerostomia (dry mouth) also results from chewing, with hypertrophy and inflammation of the salivary glands at the chewing site. In addition, this practice can lead to obvious facial asymmetry.

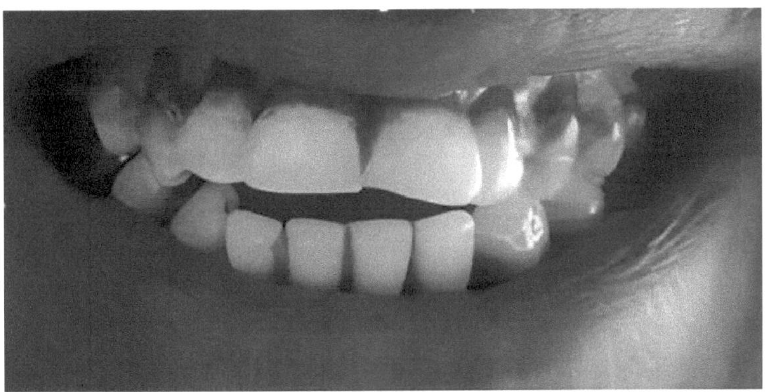

Photo 13: Teeth of a khat consumer

2.8 Khat and lung diseases

Khat consumption is associated with a variety of pulmonary disorders. Among the effects observed is an increase in tachypnea, or rapid breathing. In addition, khat users also have a higher prevalence of chronic bronchitis. These respiratory conditions are often linked to inhalation of the irritant substances contained in khat and the inflammatory response they trigger in the respiratory tract.

PART V

KHAT: MEDICINE OR DRUG?

1. Therapeutic use of Khat

The use of khat for medicinal purposes remains relatively unexplored. However, the plant is an ancient component of traditional African and Arabic medicine. Historically, it has been used to treat depression and biliary disorders. Animal studies have revealed that khat has spasmodic and analgesic effects. In addition, tests have shown that oral administration of flavonoids extracted from khat exerts anti-inflammatory activity, notably against carrageenan-induced oedema and granulomas in rats.

In Ethiopia, khat leaves and roots are traditionally used to treat ailments such as influenza, coughs, gonorrhea, asthma and other pulmonary disorders. The roots alone are also used to relieve stomach pains, while infusions of khat are sometimes administered to treat boils.

In Germany, there are preparations containing up to 5% cathine in solution, with a maximum dose of 1,600 mg per unit. These preparations are prescribed by doctors and are used as sympathomimetics.

In the UK, although khat is recognized as a medicine, it has never been used for this purpose. Recently, the antibacterial and cytotoxic properties of khat have been studied, and its active components may offer future therapeutic prospects.

In addition, some research suggests that khat's anorectic effect could lead to the development of treatments for obesity.

2. Synthetic cathinones

Synthetic cathinones are psychoactive substances which mimic the effects of cathinone, a natural compound extracted from the leaves of the khat plant (*Catha edulis*). These products are often marketed under street names such as "bath salts" or "plant fertilizers", despite their total lack of connection with these conventional products. They are also known as "designer drugs" or "synthetic stimulants".

These substances are β-keto analogues of phenylethylamines.

Some of the synthetic cathinone derivatives, such as amfepramone and pyrovalerone, have been used for their anorectic properties. However, these substances are no longer used in current medical treatments. On the other hand, bupropion, marketed under the brand name Zyban® in France, continues to be prescribed for its antidepressant and smoking cessation properties.

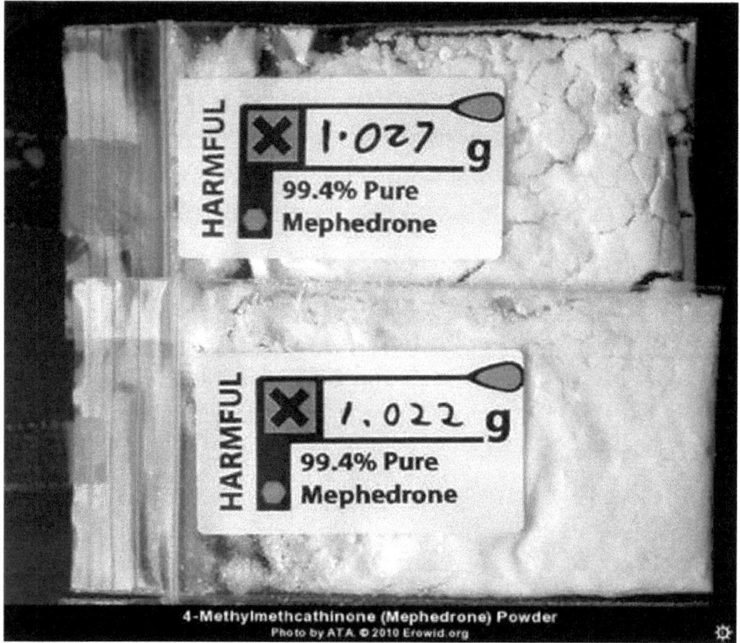

Figure 11: Mephedrone and other synthetic cathinone derivatives.

Over the past decade, substituted cathinone derivatives have emerged on European recreational drug markets. The first substance synthesized from cathinone was methcathinone. However, the most commonly encountered drugs today are mephedrone and methylone. These substances come in white or brown powder form, with a high degree of purity.

The desired effects are comparable to those of cocaine, amphetamine or MDMA (ecstasy). Consumption of mephedrone can lead to complications such as severe hyponatremia and mental confusion. In its natural state, cathinone is the S-enantiomer. However, cathinone derivatives can be present in two stereoisomeric forms, which explains why these substances are frequently found in racemic mixtures.

While some cathinones, such as mephedrone, were classified as narcotics as early as 2010, the proliferation of new, accessible and inexpensive synthetic cathinones has led the World Health Organization to include this entire family of drugs in the list of prohibited substances. Their trafficking and acquisition are now illegal, even under misleading names such as "bath salts" or "plant fertilizers".

Consumption of these substances, particularly when combined with alcohol or other narcotics, can lead to serious health effects such as palpitations, tachycardia, vomiting and headaches.

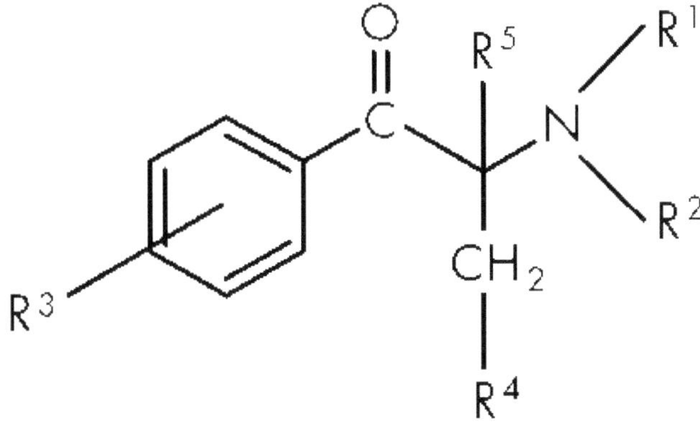

Figure 12: General structure of a cathinone derivative with substitution patterns

There are many derivatives of these compounds, as shown in Table 1.

Table 1: Structure of cathinone derivatives.

R1	R2	R3	R4	R5	Name
H	H	H	H	H	Cathinone

Methyl	H	H	H	H	Methcathinone (Ephedrone)
Methyl	Methyl	H	H	H	N,N-Dimethylcathinone (Metamfemone)
Ethyl	H	H	H	H	N-Ethylcathinone
Methyl	H	Methyl	H	H	Buphedrone
Ethyl	H	4-Methyl	H	H	4-Methyl-N-Ethylcathinone
Methyl	H	4-Methyl	H	H	Mephedrone (4-MMC ; M-CAT)
Ethyl	Ethyl	H	H	H	Amfepramone
t-Butyl	H	3-Cl	H	H	Bupropion
Methyl	H	3,4-Methylen edioxy	H	H	Methylone (βk-MDMA)
Ethyl	H	3,4-Methylen edioxy	H	H	Ethylone (βk-MDEA)
Methyl	H	4-Methyl	Methyl	H	Butylone (βk-MBDB)
Methyl	H	4-Methoxy	H	H	Methedrone (βk-PMMA)
Methyl	H	4-F	H	H	Flephedrone (4-FMC)
Methyl	H	3-F	H	H	3-Fluoromethcathinone (3-FMC)
{pyrrolidino}	H	H	H	H	α-Pyrrolidinopropiophenone (PPP)
{pyrrolidino}	4-Methyl	H	H	H	4-Methyl-α-pyrrolidinopropiophenone (MPPP)

{pyrroli dino}	4-MeO	H	H	H	4-Methoxy-α-pyrrolidinopropiophenone (MOPPP)
{pyrroli dino}	4-Methyl	Propyl	H	H	4-Methyl-α-pyrrolidino-hexanophenone (MPHP)
{pyrroli dino}	4-Methyl	Ethyl	H	H	Pyrovalerone
{pyrroli dino}	4-Methyl	Methyl	H	H	4-Methyl-α-pyrrolidino-butyrophenone (MPBP)
{pyrroli dino}	4-Methyl	H	Methyl	H	4-Methyl-α-pyrrolidino-α-methylpropiophenone
{pyrroli dino}	3,4-Methylen edioxy	H	H	H	3,4-Methylenedioxy-α-pyrrolidinopropiophenone (MDPPP)
{pyrroli dino}	3,4-Methylen edioxy	Ethyl	H	H	3,4-Methylenedioxypyrovaler one (MDPV)

3. Mephedrone

Mephedrone was first synthesized in 1929. This compound is a popular derivative of khat. It can be acquired both in physical stores and on the Internet, where its spread has been facilitated by aggressive online marketing.

Mephedrone is known by various names: Miaow, Drone, 4-MMC, MD3, Roxy, Mefedron (in Norway), Krabba (in Sweden), Meow Meow/Miaow Miaow (in the UK), Bubbles (in Scotland), Meph, Rush, Plant feeder, White Magic, and Challenge (a mixture of mephedrone and ketamine).

3.1 Structure and mechanism of action of Mephedrone

Mephedrone is a psychoactive molecule developed for research purposes. It produces stimulant and empathogenic effects similar to those of amphetamines, methamphetamine, cocaine and MDMA.

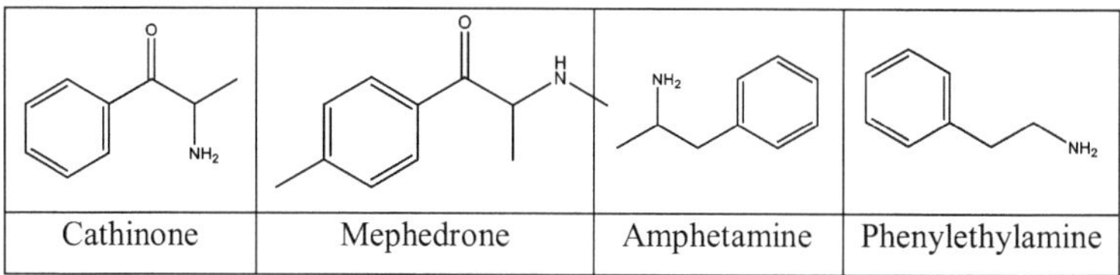

| Cathinone | Mephedrone | Amphetamine | Phenylethylamine |

Figure 13: Structure of Mephedrone and similar structures.

3.1.1 Prevalence of Mephedrone

A research project conducted by the National Addiction Center in London, involving nearly 3,000 readers of the dance magazine "Mixmag", reveals that 41.7% of those surveyed have tried mephedrone, and 33.2% have used it in the previous month. These statistics place mephedrone among the six most popular drugs among clubbers, after tobacco, alcohol, cannabis, ecstasy and cocaine.

In February 2010, a survey of 1,006 high school and university students in Scotland showed that 205 individuals (20.3%) had used mephedrone. Of these, 23.4% had taken it only once, and 4.4% on a daily basis.

Street dealers were the main source of supply in 48.8% of cases, and the Internet in 10.7%. The introduction of mephedrone in the UK appears to be correlated with a decline in the purity of ecstasy and cocaine. A university study confirmed that 20.3% of those questioned had taken mephedrone at least once; of these, almost a quarter had taken it only once, while 4.4% used it daily.

3.1.2 Structure of Méphédrone

Mephedrone comes in the form of a white to pale yellow crystalline powder. Its odor is often described as unpleasant, with notes of vanilla mixed with those of household products such as bleach, urine and electrical circuits.

Like other cathinone derivatives, mephedrone has a single chiral center, giving rise to two enantiomeric forms: (S)- and (R)-. The (S)-enantiomer of cathinone is generally more active than the (R)-, and the same is probably true of mephedrone. The synthesis of (S)-4-methylcathinone, the precursor to (S)-mephedrone, is achieved by an acylation reaction.

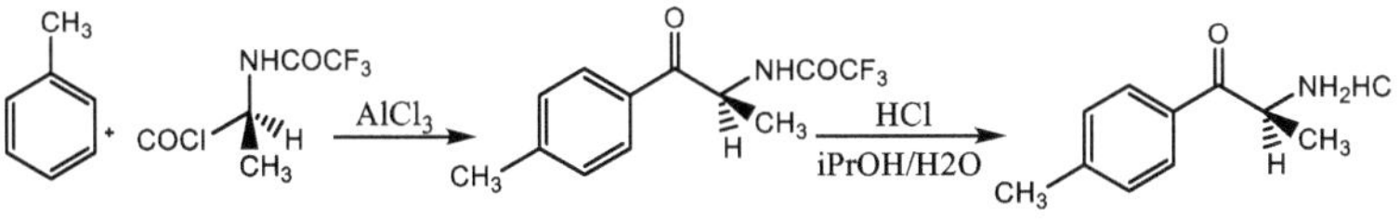

Figure 14: Stereoselective synthesis of (S)-methylcathinone

3.1.3 Pharmacology of Mephedrone

In vitro research on cathinone derivatives shows that their main mechanism of action is very similar to that of amphetamines. This mechanism is mainly characterized by an action on catecholamine transporters located at the plasma membrane. Both amphetamines and cathinone derivatives bind to noradrenaline, dopamine and serotonin transporters.

However, these derivatives have little or no effect on serotonin release or reuptake. What's more, their potential is generally lower than that of amphetamines, due to their lesser ability to cross the blood-brain barrier.

The desired effects of mephedrone include euphoria, empathy, stimulation, heightened sensory perceptions, moderate sexual stimulation, mood enhancement, reduced hostility and insecurity, and increased mental clarity and hallucinations.

Mephedrone can be detected by urine testing, usually by gas chromatography-mass spectrometry. However, it is not included in standard drug detection tests.

3.2 Trade in Mephedrone

In recent years, a number of substances imitating the effects of ecstasy and amphetamines have appeared, often referred to as "research chemicals", "legal highs", "designer drugs" or "party pills". These are available on the Internet and are generally legal, with the exception of mephedrone. One study revealed that these "legal highs" are mainly composed of mephedrone and other cathinone derivatives.

Mephedrone generally comes in the form of white or yellowish crystals or powder. It is frequently sold in powder form, but also in capsules or tablets of various colors, shapes and sizes. Each commercial form bears a different brand name and can be mixed with other substances such as caffeine, paracetamol, and sometimes with cocaine, amphetamines or even ketamine.

Pure mephedrone powder costs around €15 for 500 mg, €19 for 1 gram, €129 for 100 grams, €299 for 250 grams, and €475 for 500 grams. 250 mg tablets are sold for between £9 and £10 (€11.45 to €12.70) for two tablets.

3.3 Mephedrone administration routes

The main routes of administration for recreational use of mephedrone are insufflation (sniffing) and oral. Orally, mephedrone can be swallowed in capsule or tablet form, or by "bombing" (the process of wrapping mephedrone in cigarette paper and swallowing it). Thanks to its solubility in water, it can also be taken rectally (as an enema or gelatin capsule) and injected intravenously, although these methods are less common.

Some users choose to smoke mephedrone, but this practice is in the minority. The oral route is often preferred for its longer duration of effect, which can range from 2 to 4 hours, with fewer side effects and a reduced need to re-dose.

Insufflation, on the other hand, is the most widely used method, as it produces effects in a matter of minutes, with a peak effect reached in less than thirty minutes and a rapid "descent". Recommended doses for insufflation generally range from 25 to 75 mg. The effect can be influenced by diet, hence the advice to consume on an empty stomach.

Some users combine insufflation and oral intake to benefit from both a rapid and prolonged effect. Injection, whether intramuscular or intravenous, requires a lower dose than oral intake, and can be combined with other substances such as heroin or ketamine.

3.4 Toxicity of Mephedrone

When used recreationally, mephedrone is often combined with other psychoactive substances such as alcohol, methylone, cocaine, ketamine, ecstasy, heroin and cannabis, as well as pharmaceutical stimulants like modafinil and viagra. This substance affects various bodily systems: it influences thermoregulation and locomotor activity, leading to a drop in body temperature after a sudden rise, as well as an increase in physical activity. Muscular effects such as tense cheeks, moderate muscle tension and bruxism are also observed.

Cardiovascularly, mephedrone can cause tachycardia, elevated blood pressure and chest pain, as well as peripheral vasoconstriction leading to vascular problems. Rare cases of hepatotoxicity and nephrotoxicity have been reported.

Adverse reactions observed include:

- ✓ Difficulty breathing ;
- ✓ Dehydration and dry mouth;
- ✓ Nausea, vomiting, digestive discomfort and abdominal pain;
- ✓ Flu-like symptoms ;
- ✓ Skin rash, itching, pustules ;
- ✓ Numbness and loss of sensitivity;
- ✓ Joint pain ;
- ✓ Discoloration of extremities and joints;
- ✓ Mydriasis ;

- ✓ Nystagmus ;
- ✓ Nasal pain after insufflation, with clots and mucus the next day, as well as burning and ulceration of the mouth;
- ✓ Dizziness and vertigo ;
- ✓ Sudden cervicogenic headaches ;
- ✓ Tremors and convulsions ;
- ✓ Drunkenness;
- ✓ Increased sexual desire ;
- ✓ Nightmares ;
- ✓ Insomnia ;
- ✓ Agitation;
- ✓ Fatigue ;
- ✓ Loss of concentration, memory problems and amnesia;
- ✓ Anxiety ;
- ✓ Paranoia;
- ✓ Depression;
- ✓ Hallucinations.

Users report that adverse effects increase exponentially with intensive use. Although a few deaths have been reported, mephedrone has not always been identified as the direct cause; in most cases, its use was associated with heroin or other drugs. One notable case of extreme agitation followed by death was observed in the Netherlands. In the event of mephedrone intoxication, management consists of a period of observation and administration of rehydration solutions, plus benzodiazepines if necessary.

3.5 Legal status of Méphédrone

Regulations concerning mephedrone were introduced following reports of the health and social risks associated with the substance, as well as the rise in organized crime linked to its use. Mephedrone is controlled in Germany, Sweden, Norway, Denmark, Estonia, Romania, Ireland and Israel. In Finland and the Netherlands, it is classified as a medicinal substance. In France, it was classified as a narcotic by a decree published in the Journal Officiel on June 11, 2010. In the UK, mephedrone and other cathinone derivatives have been classified as class B narcotics since April 16, 2010.

PART VI
KHAT: LEGISLATION, ECONOMICS AND SOCIETY

1. Socio-economic impact of Khat

1.1 Situation of Khat in the countries of origin

In Yemen, over 80% of men consume khat, often for more than four hours a day, while 50% of women are also involved. These figures may be underestimated, particularly in view of the growing consumption among young women. A recent study indicates that 73% of women in Yemen chew khat on a regular or occasional basis, and it is estimated that 15-20% of children under the age of 12 are also daily consumers. This practice has a significant impact on users' social and economic lives.

Jamal Al-Shammi, president of the Yemeni NGO Democratic School, points out that the harmful effects of khat go beyond those on individual health. Khat disrupts family life by isolating family members from each other: parents chew separately, and children are left to their own devices. An anonymous testimonial from a middle-class woman who has been using khat for over 15 years reveals that her husband would leave to chew after dinner and wouldn't return until midnight. For her, these sessions were a way of escaping the daily monotony, leaving her children under the supervision of a maid. Similarly, a sociological study revealed that khat is a major cause of divorce in Djibouti.

Khat consumption also generates social pressure. People who do not consume khat may suffer a form of social exclusion. In addition, the time devoted to khat negatively affects productivity: people may lose their dignity by accepting bribes to finance their consumption.

Some families prioritize the purchase of khat to the detriment of essential food needs for their children, reserving more than 50% of their income for this expense. This situation affects not only family finances, but also the national economy.

Dr. M. Taghi Yasamy, Regional Administrator for Mental Health and Substance Abuse in the Mediterranean Region, notes that khat consumption causes insomnia, leads to late awakening and thus reduces working time and performance, with significant economic repercussions.

As a result, khat consumption leads to fewer hours worked and lower incomes, increasing the risk of malnutrition in families. In addition, studies show that khat-related habits reduce productivity in countries such as Ethiopia, Somalia, Uganda, Kenya and Djibouti.

To counter the effects of khat cultivation, the Ethiopian authorities have encouraged the substitution of khat crops with alternative crops such as coffee.

1.2 Situation of Khat in Western countries

The expansion of khat consumption in Western countries is linked to improved means of transport and an increase in the number of migrants from khat-producing regions. In Australia, for example, khat imports rose from 70 kg in 1997 to 18,830 kg in 2007. This increase coincides with a 187% rise in the population originating from sub-Saharan Africa. It is therefore crucial to monitor this trend closely.

In the Netherlands, studies on the social impact of khat indicate that the substance is not associated with incidents of problem behavior, the development of organized crime or significant negative health effects.

In the UK, khat consumption is considered to have a relatively minor impact compared with alcohol. However, more detailed research shows that khat is imported by air, mainly via Heathrow airport, from Ethiopia, Kenya and Yemen. In 2010, around 57.7 tonnes of khat per week were imported, although this figure may be underestimated due to under-reporting. It should be noted that some of this khat is destined for other markets in Europe and the USA. Since the 1990s, imports have increased considerably, from less than 7 tonnes per week at the time.

In the UK, khat consumption is mainly observed among migrants from East Africa and the Red Sea coast, including Somalis, Ethiopians, Kenyans and Yemenis. For these communities, khat plays an important social role, facilitating community cohesion and business relationships.

Khat-drinking sessions strengthen social ties and can help to establish professional networks, making it easier to find a job. The practice is perceived as a cultural pastime of little consequence.

Nevertheless, khat consumption can lead to social problems. The time spent chewing and recovering after sessions can limit employment opportunities in Somali, Ethiopian and Yemeni communities. Sleep disturbances linked to khat consumption can also be a barrier to employment. The links between khat and crime are weak, with public order disturbances generally limited to behaviors such as spitting out pieces of khat and organizing street meetings. However, there are associations between khat consumption and cases of violence, particularly domestic and marital violence, as well as consumption-induced psychosis. In Denmark, for example, it is reported that two-thirds of male khat abusers are divorced.

Finally, the high proportion of income spent on khat in low-income households is a cause for concern, especially among women. The integration of migrants into host societies is affected by various factors, including language, which is a more significant barrier than khat consumption.

2. Legal aspects of Khat

Khat is widely consumed in the regions surrounding the Red Sea, such as Ethiopia, Kenya, Tanzania, Somalia and Djibouti; but the practice reaches its highest level in Yemen. In many Islamic countries such as Yemen, Somalia and Djibouti, as well as in Muslim communities in Ethiopia and Kenya, khat is legal, making it a substance of choice. Unlike alcohol, khat is not explicitly forbidden by religious institutions in these regions.

However, some Muslim countries, such as Saudi Arabia, prohibit khat for religious and economic reasons. Penalties for possession and use of khat are similar to those applicable to opium or cannabis.

In Somalia, the Union of Islamic Courts, which took power in 2006, banned khat, although its use persists in the south of the country despite this ban.

In Djibouti, Somaliland and Yemen, as well as in Uganda, khat is legal, although the legislation in Uganda is often unclear. In Eritrea and Tanzania, on the other hand, khat is illegal.

The difference in treatment of khat compared to other substances such as opium or cannabis can be explained by the fact that khat does not cause significant antisocial behavior. It is often compared to amphetamines or caffeine in terms of effects. In Ethiopia and neighboring countries, khat consumption is seen as a social gathering, similar to alcohol consumption in Western countries.

In Europe, khat is generally viewed with suspicion. This attitude is partly due to its unattractive mode of use and relatively low potency compared to amphetamines. However, it is important to develop preventive measures to avoid the spread of this habit, particularly among immigrant populations. In the UK, for example, khat is legal, and markets and distribution networks are well established. Some people in Europe, including expatriates living in Yemen, have also adopted the practice.

Khat is listed as a narcotic in some countries, such as France, where it was classified by ministerial decree on February 20, 1957. In Europe, legislation on khat varies considerably: it is banned in Switzerland, Sweden, Denmark, Italy, Germany and Finland, but tolerated in the UK and the Netherlands.

Outside Europe, khat is illegal in the USA, Australia, New Zealand, Norway and Canada. On the other hand, Cyprus, the Czech Republic, Greece, Malta, the Netherlands, the UK, Spain and Portugal have no specific legislation on khat.

Figure 15: Legal status of khat in EU member states and Norway, and seizure statistics

Note: Hungary does not control khat, but has data on seizures due to cathinone control.

Conclusion

Khat, a plant endemic to East Africa, is traditionally consumed during long chewing sessions. Its cultivation is concentrated in the mountainous regions of the highlands, offering ideal conditions for its growth. Once harvested, khat is transported to the cities, where it is the object of a lucrative trade.

The alkaloids present in khat have euphoric and stimulating effects, which are much sought-after by consumers. However, this plant is not without undesirable effects. Consumption of khat leads to short-term effects such as insomnia, lethargy, mild tremors and depression, which can develop into serious health problems in the medium and long term for regular users. Although the direct relationship between khat use and certain serious illnesses is not always clearly established, it is difficult to completely rule out the role of khat in the onset of these illnesses.

Research has explored khat alkaloids for the development of pharmaceuticals. However, many khat-derived treatments, such as anorectics, have become obsolete. Currently, only one smoking cessation drug remains relevant, but it is considered a second- or third-line treatment. Nevertheless, the recently discovered cytotoxic properties of khat are arousing growing interest in its potential as a chemotherapy agent. At the same time, synthetic molecules derived from khat, such as mephedrone, have emerged. These synthetic drugs are spreading in some European countries such as the UK and the Netherlands, not least because of the declining quality of other drugs and their relatively low price, making them popular with young people.

Legislation on khat varies considerably around the world. In the Arab world, Europe and America, each country applies its own rules. France has had khat on its list of narcotics since 1957. In Yemen, khat consumption has a significant impact on social and economic life, particularly at family level.

In Europe, khat is mainly introduced by migrants from regions where it is commonly consumed. The UK, because of its legislation and colonial past, is particularly affected by this import.

Although the way khat is consumed is not very attractive in temperate climates, and its physical effects, such as tooth discoloration, may deter some. The euphoric effect of khat remains attractive to some individuals. It is therefore important to keep an eye on synthetic drugs derived from khat. These substances, which are difficult to detect and often taken in dissolved or snorted form, are becoming increasingly popular and deserve special attention.

The debate on khat oscillates between perceptions of its use as a simple social stimulant and more serious concerns about its effects on mental and physical health. For some, khat is perceived as a substance with mild hallucinogenic effects, fuelling altered states of consciousness that can lead to episodes of disconnection from reality, and even hallucinations in chronic users. These effects, while marginal for some users, raise questions about long-term psychological safety.

On the other hand, recent studies are exploring the therapeutic potential of khat, particularly in relation to its cytotoxic properties, which could open up new avenues for the treatment of certain forms of cancer. However, these therapeutic hypotheses remain in the embryonic stage and require in-depth research to distinguish the true potential benefits from the risks inherent in consuming this plant. Thus, the mysteries surrounding khat, oscillating between hallucination and therapy, reflect the complexity of this multifaceted plant, whose impact on public health continues to be a subject of intense debate.

BIBLIOGRAPHICAL REFERENCES

1. Mohamed Abdoul-Latif, F., Ainane, A., Merito, A., Houmed Aboubaker, I., Mohamed, H., Cherroud, S., & Ainane, T. (2024). The Effects of Khat Chewing among Djiboutians: Dental Chemical Studies, Gingival Histopathological Analyses and Bioinformatics Approaches. *Bioengineering*, *11* (7), 716.

2. Mohamed Abdoul-Latif, F., Ainane, A., Houmed Aboubaker, I., Merito Ali, A., El Montassir, Z., Kciuk, M., ... & Ainane, T. (2023). Chemical composition of the essential oil of *Catha edulis* Forsk from Djibouti and its toxicological investigations *in vivo* and *in vitro*. *Processes*, *11* (5), 1324.

3. Ayano, G., Ayalew, M., Bedaso, A., & Duko, B. (2024). Epidemiology of Khat (*Catha edulis*) chewing in Ethiopia: a systematic review and meta-analysis. *Journal of psychoactive drugs*, *56* (1), 40-49.

4. Wood, E. A., Case, S. J., Collins, S. L., Stark, H., & Wilfong, T. (2024). From traditional to transactional: exploration of khat use in Ethiopia through an interpretative phenomenological analysis. *BMC Public Health*, *24* (1), 1887.

5. Ademe, B. W., Brimer, L., Dalsgaard, A., & Belachew, T. (2020). Chemical and microbiological hazards of Khat (*Catha edulis*) from field to chewing in Ethiopia. *GSC Biological and Pharmaceutical Sciences*, *11* (1), 024-035.

6. Omare, M. O. (2020). Contemporary trends in the use of khat for recreational purposes and its possible health implications. *Open Access Library Journal*, *7* (12), 1.

7. Weyesa, G. W. (2021). Practices and challenges of pestiside application on Khat Farm, the case of Kersa Woreda, Jimma Zone, south West Ethiopia. *Sci Dev*, *2* (1), 7-14.

8. Szendrei, K. (1980). The chemistry of khat. *Bull Narc*, *32* (3), 5-35.

9. Getasetegn, M. (2016). Chemical composition of *Catha edulis* (khat): a review. *Phytochemistry Reviews*, *15*, 907-920.

10. Armstrong, E. G. (2008). Research note: Crime, chemicals, and culture: On the complexity of khat. *Journal of Drug Issues*, *38* (2), 631-648.

11. Ademe, B. W., Brimer, L., Dalsgaard, A., & Belachew, T. (2020). Chemical and microbiological hazards of Khat (*Catha edulis*) from field to chewing in Ethiopia. *GSC Biological and Pharmaceutical Sciences*, *11* (1), 024-035.

12. Engidawork, E. (2017). Pharmacological and toxicological effects of *Catha edulis* F. (Khat). *Phytotherapy Research*, *31* (7), 1019-1028.

13. Valente, M. J., Guedes de Pinho, P., de Lourdes Bastos, M., Carvalho, F., & Carvalho, M. (2014). Khat and synthetic cathinones: a review. *Archives of toxicology*, *88*, 15-45.

14. Basker, G. V. (2013). A review on hazards of khat chewing. *Int J Pharm Sci,*
    5 (3), 74-77.

APPENDICES

Appendix 1: Botanical glossary

Appendix 2: APG classification

Appendix 3: Comparison of chemical structures of khat alkaloids with amphetamine and ecstasy (MDMA) structures

Appendix 4: Khat alkaloids

Appendix 5: Khat cathedulins

Appendix 6: Triterpene quinones

Appendix 1: Botanical glossary

Acuminate (n.m.) (acuminate)

Refers to an organ, usually a leaf or petiole, whose tip ends abruptly in a fine, more or less elongated point. For example, the leaves of birch (Betula), pear (Pyrus) and lime (Tilia) often have this characteristic.

Albumen (n.m.) (albumen, endosperm)

Reserve tissue found in Angiosperm seeds, located around the embryo. It plays a crucial role in providing the nutrients required for the embryo's initial development. This tissue is formed by the fusion of one of the male gametes with the two central nuclei of the embryo sac, forming an accessory zygote with a variable level of ploidy. It then divides to create either a syncytium (nuclear albumen) or a cellular tissue (cellular albumen). In albuminized seeds, such as corn and castor, the albumen persists, while in exalbuminized seeds, such as beans and peas, it is temporary. Chemical reserves in the albumen vary: starch (cereals), fats (coconuts, castor beans), hemicelluloses (dates), proteins (peas, castor beans).

Anther (n.f.) (anther)

Terminal and generally enlarged part of a stamen, containing pollen grains in two pollen lodges, each formed by the fusion of two pollen sacs (equivalent to microspores). The diploid mother cells in these pollen sacs undergo meiosis to form haploid microspores, which evolve into pollen grains (male gametophytes).

Apex (n.m.) (apex)

Terminal part of a plant organ, such as a stem or root, containing the apical meristem, responsible for the organ's growth in length. The apex is crucial to the growth and development of plant organs.

Arille (n.m.) (aril)

A tegumentary outgrowth that develops around certain mature seeds, but does not adhere to the outer tegument. The aril arises from the base of the funicle at the hilum.

It is often fleshy and can be brightly colored, such as red, sometimes almost completely covering the seed. An example is the aril of Yew (*Taxus baccata*).

Bract (n.f.) (bract)

Small leafy or membranous structure, often colored differently from the leaves, located in the axils of flowers or inflorescences in some species. Bracts are neither sepals nor petals. They can be :

- ✓ **Shape**: They can be simple or highly developed, reaching large sizes.
- ✓ **Examples**:
    - ➢ The artichoke's edible "leaves" are bracts.
    - ➢ The colorful bracts of *Bougainvillea* flowers are highly decorative.
    - ➢ In some cases, they form a **collar** around the flower.
    - ➢ The large, leathery bracts are called **spathes**, like those on corn.

Caduc (deciduous) (adj.) (caducous, deciduous)

Refers to an organ (leaf, sepal, petal, etc.) that dies and detaches after fulfilling its function during each annual life cycle. Deciduous organs are renewed each season:

- ✓ **Examples**:
    - ➢ Chestnut (*Castanea*) leaves fall in autumn.
    - ➢ The sepals of the poppy (*Papaver*) detach after flowering.
    - ➢ The leaves of maple (*Acer*) and beech (*Fagus*) are also deciduous.

Synonym: Decidu

Antonym: Persistent

Calyx (n.m.) (calyx, pl. calyces)

The outer covering of the flower, usually green, made up of all the sepals, which covers the base of the corolla. It can :

- ✓ **Type of Sepals :**
    - ➢ **Free**: Not fused (e.g. Wallflower, Ranunculus).
    - ➢ **Welded**: Partially fused (e.g. Mint, Carnation).
    - ➢ **Petalloids**: Colored like petals (e.g. Tulip, Garlic).
- ✓ **Features :**

> **Possibly** persistent (remaining attached after flowering, e.g. apple tree) or **deciduous** (detaching after flowering, e.g. poppy).
> Possibly **accrescent** (growing after fertilization).

Capsule (n.e.) (capsule, seed case)

A dehiscent dry fruit in Angiosperms, formed from a syncarpous gynoecium, which opens to release the seeds. Capsules can open in different ways:

- ✓ **Types of dehiscence** :
    - ➤ **Par Valves**: Sections that separate to release the seeds (e.g.: Poppy - *Papaver somniferum*).
    - ➤ **Pores**: Small orifices (e.g. Gentian - *Gentiana* sp.).
    - ➤ **By Teeth or Lid**: Some capsules are opened by teeth or a lid.

Cotyledon (n.m.) (cotyledon)

Primordial leaf(s) present in the embryo of Spermaphytes, playing a crucial role during germination:

- ✓ **Monocotyledons**: have a single cotyledon (e.g. Orchidaceae, Liliaceae, Poaceae, Arecaceae).
- ✓ **Dicotyledons**: have two cotyledons (e.g. Rosaceae, Fabaceae, Solanaceae).
- ✓ **Function**:
    - ➤ **Albuminous seeds** : Cotyledons are leafy (e.g. castor oil) and store reserves.
    - ➤ **Exalbuminous seeds** : Cotyledons accumulate reserve substances (e.g. beans).
- ✓ **Germination** :
    - ➤ **Epigeous** : The cotyledons are carried out of the soil and carry out photosynthesis (e.g. Beans, Beech).
    - ➤ **Hypogea** : The cotyledons remain in the soil, providing nutrient reserves (e.g. Broad Bean, Oak).
- ✓ **Gymnosperms**: Can have up to a dozen cotyledons.

Cyme (n.f.) (cyme)

A type of inflorescence in which a main axis bears a terminal flower, with secondary axes branching out on either side in the same way:

- ✓ **Uniparous cyme**: Lateral axes develop on one side only. Two subtypes:
  - ➢ **Scorpioid cyme**: the lateral axes always appear on the same side, like an unrolled crosier (e.g. *Solanum tuberosum*).
  - ➢ **Helical cycle**: the lateral axes appear alternately on each side.
- ✓ **Biparous cyme**: Lateral axes are opposite (e.g.: Linum - *Linum*, Soapwort - *Saponaria*).
- ✓ **Cyme Multipare**: several lateral axes start from the same point on the main axis.

Dehiscence (adj.) (dehiscent)

Qualifies an organ (fruit, anther, sporangium) that opens by itself at maturity to release its contents (seeds, pollen, spores):

- ✓ **Dehiscence modes** :
  - ➢ **Opercular**: By an operculum.
  - ➢ **Apical or Poricidal**: Through pores at the top (e.g. certain capsules).
  - ➢ **Loculicide**: Opening in the back of the carpel lodges (e.g. Pansy, Tulip).
  - ➢ **Septicide** : Along the septa between loges (e.g.: Colchicum, Gentian).
  - ➢ **Septifrage**: Separation along the septa, but the fruit remains attached.

Dichasial (adj.) (dichasial)

Relating to a **dichasium** inflorescence:

- ✓ **Dichasium**: An inflorescence where each floral axis divides into two lateral branches, each bearing a terminal flower.
- ✓ **Double branching**: From a terminal bud, the inflorescence branches into two axes, each continuing to branch in the same way.
- ✓ **Example**: Dichasial cymes are typical of certain species such as **Gypsophila**.

Disque (n.m.) (disc, disk)

Glandular organ, often flattened, located around the pistil in certain Dicotyledonous flowers :

✓ **Function**: Nectar production, which attracts pollinators.

✓ **Position**: Above the floral receptacle and below the stamens and pistil.

✓ **Example: Rutaceae**, such as lemons and oranges.

✓ **Intrastaminal disc**: Disc located inside the stamens, often in the center of flowers.

Drupe (n.f.) (drupe)

Indehiscent fruit with a distinct structure:

✓ **Epicarp**: Membranous, the outer shell of the fruit.

✓ **Mesocarp**: Hinged and pulpy, the edible part of the fruit.

✓ **Endocarp**: Sclerified and hard, forming the core surrounding the seed.

✓ **Examples**:

   ➢ **Apricot** (*Prunus armeniaca*)

   ➢ **Cherry** (*Prunus cerasus*)

   ➢ **Peach** (*Prunus persica*)

   ➢ **Olive** (*Olea europaea*)

✓ **Compound drupe**: Formed from several drupes, called **polydrupes** (e.g. raspberry).

Erect (adj.) (upright)

Qualifies a plant or part of a plant that is erect or tall, standing vertically:

✓ **Example**: Erect stems of certain plants such as the **Bumblebee** or the **Suns**.

Etamine (n.f.) (stamen)

Male organ of a flower, formed by two main parts:

✓ **Filet**: Elongated, thin part supporting the anther.

✓ **Anther**: swollen terminal part where pollen grains are produced in pollen sacs (microsporangia), often organized into two lodges separated by a connective.

✓ **Number of stamens**: Varies according to species, ranging from one (e.g. Orchis) to several dozen (e.g. Ranunculus).

✓ **Type of Flowers** :

   ➢ **Staminate flowers**: Have only stamens and no pistil.

Extrorse (adj.) (extrorse)

Refers to the anthers of the stamens which open towards the outside of the flower, allowing pollen to be dispersed outwards.

- ✓ **Examples: Anemone** anthers (Renonculaceae), *Papaver rhoeas* (Papaveraceae).

Fascicular (adj.) (fascicular, fascicled)

Describes similar organs arranged in bundles, i.e. grouped by their extremities at a single common point.

- ✓ **Example: Poaceae** (grass) roots, which can emerge in a bundle from a central point.

Fimbrié (adj.) (fimbriated)

Refers to the margin of an organ, such as a leaf or petal, when it is finely cut into single elements or fimbriae, resembling a fringe.

- ✓ **Example: Carnation** petals (*Dianthus superbus*).

Glabrous (adj.) (glabrous, smooth)

Refers to an organ with a smooth surface, free of hairs or growths.

- ✓ **Examples: Boxwood** leaves (*Buxus sempervirens*), **Cabbage** (*Brassica oleracea*), **Licorice** (*Glycyrrhiza glabra*).
- ✓ **Antonym**: Pubescent, hairy.

Inferior (adj.) (inferior)

An ovary or fruit located below the level of insertion of the other floral parts (petals, sepals, stamens).

- ✓ **Examples: Apiaceae** ovaries (e.g. carrots), **Orchidaceae, Currants** (Ribes).

Inflorescence (n.f.) (inflorescence)

A group of flowers forming an organized and often morphologically distinct whole.

Introrse (adj.) (introrse)

Said of a stamen whose anther opens towards the inside of the flower.

- ✓ **Example**: *Campanula* anthers.

Latrorse (adj.) (latrorse)

Said of a stamen whose anther opens sideways, often towards other anthers rather than outwards.

- ✓ **Example**: Some **Fuchsia** anthers.

Orthotropic (adj.) (orthotropous)

1. **Orthotropic ovule**: An ovule in which the hilum, chalaza and micropyle are vertically aligned, with the micropyle directly opposite the funiculus.
   - ➤ **Example**: Many **Gymnosperms**, **nettle** ovule (*Urtica*).
2. By extension, an almost vertical or straight axis.

Panicle (n.f.) (panicle)

Type of indefinite inflorescence characterized by a cluster of clusters, with several flowers on each stalk. Often pyramidal or conical in shape.

- ✓ **Examples**: Male inflorescence of **corn** (*Zea mays*), **oats** (*Avena*).

Pédoncule (n.m.) (peduncle)

Flower or fruit-bearing axis. In the context of an inflorescence, the peduncle is the main axis, while the pedicels are the finer axes carrying the individual flowers.

- ✓ Stalkless flowers and fruits are called **sessile**.

Penné, Pennati- (or Pinnati-) (adj.)

Describes venation in which the secondary veins are regularly arranged on either side of a main vein.

- ✓ **Example**: **Pennatinervated** sheet.

Pericarp (n.m.) (pericarp)

Wall of a ripe fruit, generally divided into three layers:

- ✓ **Epicarp**: External, often thin.

- ✓ **Mesocarp**: Intermediate, fleshy and edible in certain fruits.
- ✓ **Endocarp**: Internal, often hard, containing the seed.

Petal (n.m.) (petal)

The colorful and often variegated part of the perianth that forms the corolla and plays a role in attracting pollinators.

- ✓ Petals can be :
  - ➤ **Free** (dialypetal), as in the **Apple tree**.
  - ➤ **Welded** (gamopetal), as in **Borage**.
  - ➤ **Absent** (apetals), as in **Nettle**.

Petiole (n.m.) (petiole, leafstalk)

The narrowed part of the leaf that connects the blade to the stem. Contains the conductive tissues that irrigate the leaf blade.

- ✓ Stalkless leaves are called **sessile**.

Plagiotrope (adj.) (plagiotropic)

1. Refers to branches or plant parts with an oblique orientation, often close to the horizontal, resulting in differentiation between upper and lower surfaces.
   - ➤ **Example: Cedar of Lebanon** (*Cedrus libani*), *Araucaria excelsa*.
2. Also used to describe rhizomes that grow horizontally.

Raceme (n.m.) (raceme)

A type of inflorescence where flowers are borne on pedicels of varying lengths along a main axis.

- ✓ **Example: Raceme** of certain **Liliaceae**.

Reticulate (adj.) (netted, reticulate(d))

1. Describes venation in which the veins are interlaced like a network.
   - ➤ **Example:** venation in **dicots**.
2. Can also refer to the lignification of certain vessels, forming a network.

Samare (n.f.) (samara)

Dry, indehiscent fruit with a membranous wing developed from the epicarp, facilitating wind dispersal.

- ✓ **Examples**: **Maple** (*Acer pseudoplatanus*), **Ash** (*Fraxinus*), **Elm** (*Ulmus*).

Schizocarp (n.m.) (schizocarp)

Dry, indehiscent fruit from an ovary composed of several fused carpels. At maturity, the loges transform into achenes, which separate from each other but remain attached to a common axis called the carpophore. Each achene is a mericarp.

- ✓ **Examples**: Fruits of **Malvaceae** (such as hibiscus), **Geraniaceae** (such as geranium).

Sempervirent (adj.) (sempervirent, evergreen)

Describes a tree or shrub whose leaves remain green over several growing seasons. The foliage is evergreen, although individual leaves may be deciduous. Defoliation is spread out over time, masked by the continuous growth of new leaves.

- ✓ **Example**: Most **conifers** (e.g. pine, spruce).
- ✓ **Antonym**: Caducifoliate (or deciduous).

Sepal (n.m.) (sepal)

Component part of the flower, generally green in color, located outside the corolla and together forming the calyx. Sepals can be :

- ✓ **Free from** each other (dialypetal calyx),
- ✓ **Welded** together (gamosepal calyx),
- ✓ **Absent** in some cases (as in **wheat**).
- ✓ Sometimes, sepals can resemble petals (tepaloids) or be identical to petals (tepals).

Serrated (adj.) (serrate)

Said of an organ with serrated, sawtooth-like margins.

✓ **Example**: The edges of certain leaves, such as **nettle**.

Stigma (n.m.) (stigma)

Sticky, glandular end of the style or apical part of the pistil, suitable for receiving pollen grains.

✓ **Example**: **Poppy** stigma (*Papaver rhoeas*), **Tulip** (*Tulipa*).

Stipule (n.f.) (stipule)

A foliaceous or scaly appendage, sometimes spiny or glandular, located in pairs at the base of the leaf petiole of certain species. Stipules may be deciduous or evergreen.

✓ **Examples**: Stipules of **Pea** (*Pisum sativum*), **Bramble** (*Rubus*), **Clover** (*Trifolium*).

✓ In some cases, stipules can be highly developed or replace leaves transformed into spines (as in ***Galium***).

Style (n.m.) (style)

Elongated, tapering part of the pistil extending the ovary, with the stigma at its tip. The style may be absent in some flowers, with the stigma attached directly to the ovary.

✓ **Examples**: Style of **poppies** (*Papaver rhoeas*), **Tulips** (*Tulipa*).

Super (adj.) (superior)

Qualifies an ovary whose insertion is above the level of the other floral parts (sepal, petal, stamen). In this case, the flower is said to be hypogynous or superovariate.

✓ **Examples**: *Ranunculus*, *Lycopersicon esculentum*.

✓ **Antonym**: Infere.

Tegument (n.m.) (tegument, integument)

1. Protective covering of an organ, such as an ovule, seed or fruit.
2. Envelope surrounding the ovule, developing into the testa or seed coat at maturity. It can have a variety of textures and colors, with outgrowths facilitating dissemination.

> ➤ **Examples**: **Dandelion** (*Taraxacum*) egret, **pine** (*Pinus*) wing.

Thyrse (n.f.) (thyrse, thyrsus)

A type of compound inflorescence where the cymes are grouped together in a cluster. The general appearance is that of a pyramidal or ovoid cluster.

- ✓ **Examples**: **Lilac** thyrse (*Syringa*), **privet** (*Ligustrum*).

Verticille (n.m.) (verticil, whorl)

A group of similar organs (leaves, sepals, petals, stamens) arranged in a circle at the same level around or at the top of an axis. There are generally at least three elements in a whorl.

- ✓ **Examples**: **Common Juniper** (*Juniperus communis*) needles, *Erica verticillata* (Ericaceae) whorled leaves in threes, and **Oleander** (*Nerium oleander*, Apocynaceae).
- ✓ **Note**: The term whorl can be applied to a variety of anatomical elements, while the term cycle applies mainly to floral parts.

Appendix 2: APG classification.

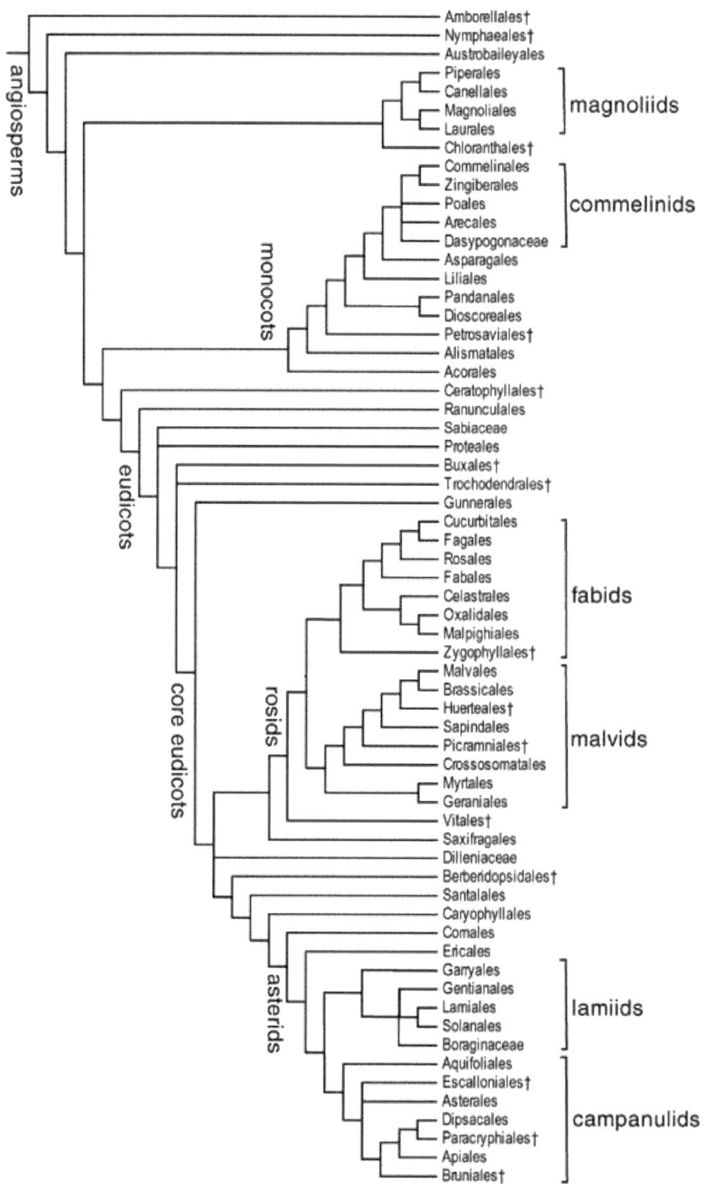

Appendix 3: Comparison of the chemical structures of khat alkaloids with those of amphetamine and ecstasy (MDMA).

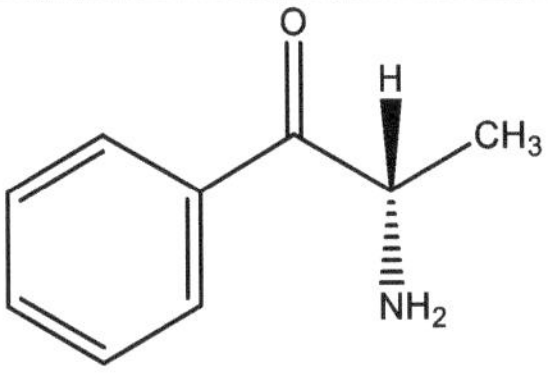

S-(-) Cathinone

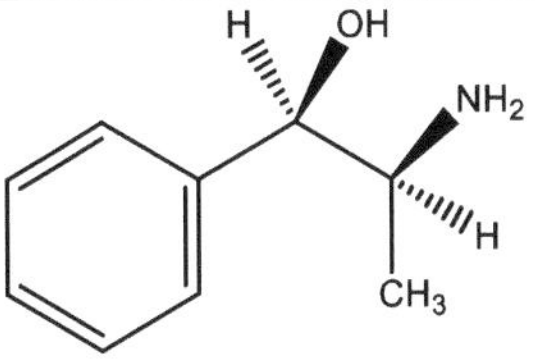

1R, 2S(-) Norephedrine

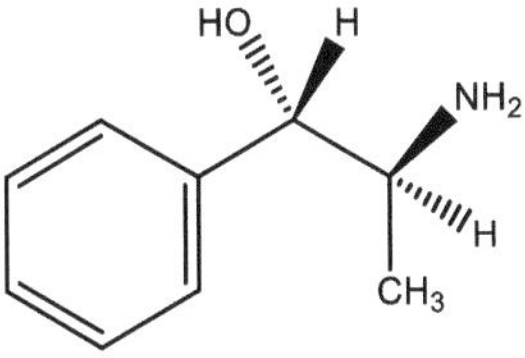

1S, 2S(+) Norpseudoephedrine

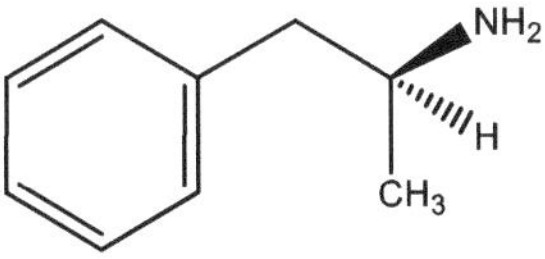

S-(+)- Amphetamine

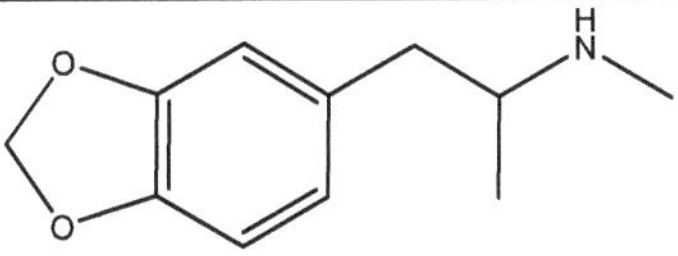

Ecstasy, MDMA

Appendix 4: Khat alkaloids.

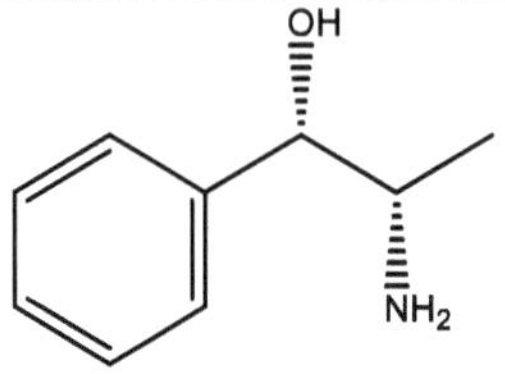

OH
NH2

Cathine

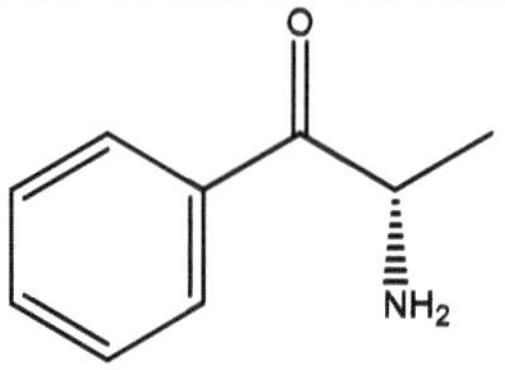

O
NH2

Cathinone

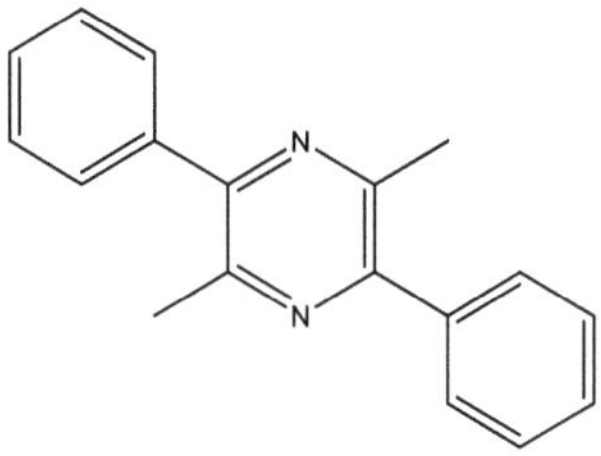

N
N

2,5-Dimethyl-3,6-diphenylpyrazine

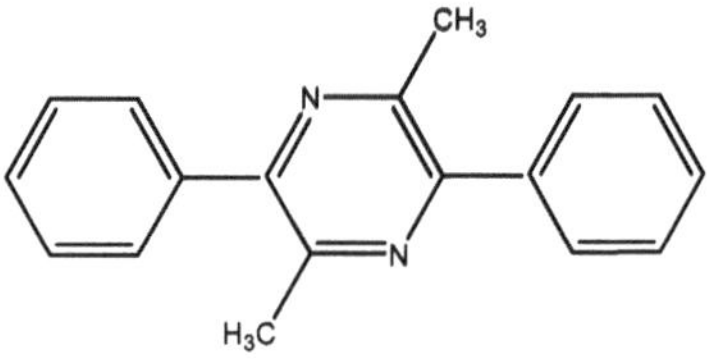

CH3
N
N
H3C

Dimère de cathinone

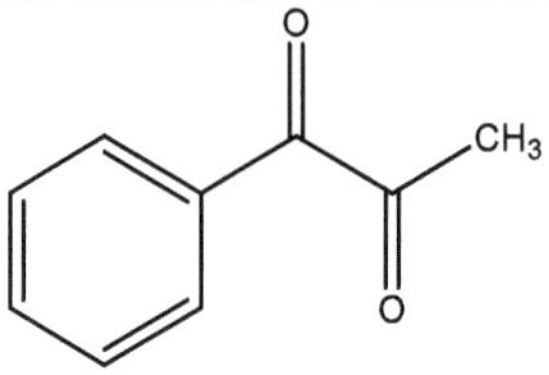

O
CH3
O

1-phenylpropane-1,2-dione

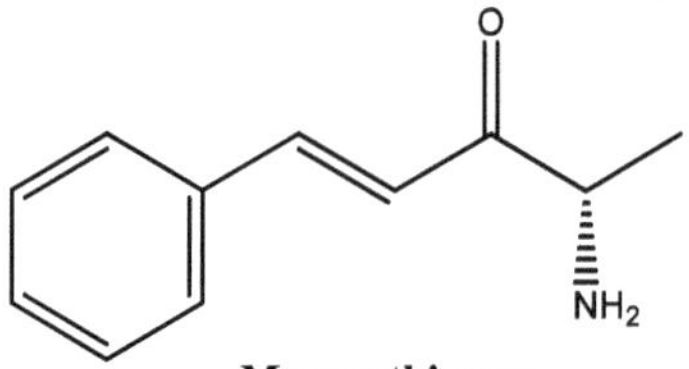

O
NH2

Merucathinone

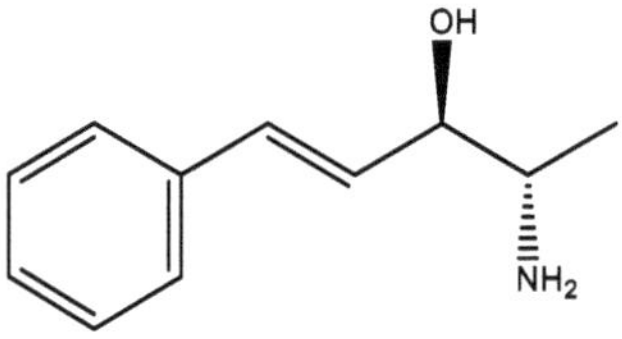

OH
NH2

Merucathine

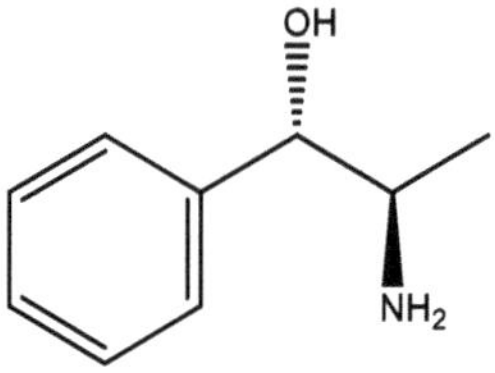

OH
NH2

Noréphédrine

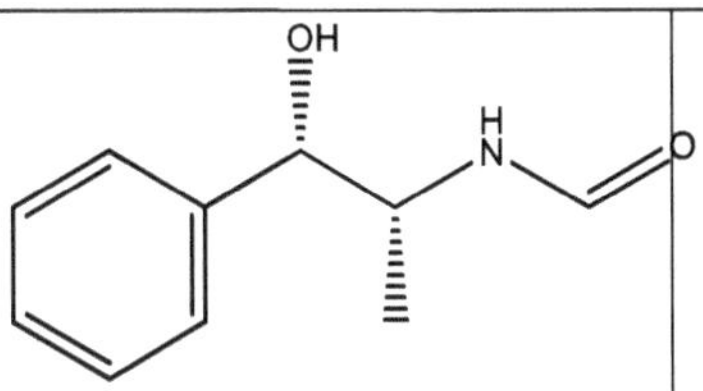

OH
H
N
O
N-formylnoréphédrine

Appendix 5: Khat cathedulines.

Low molecular weight cathedulins :

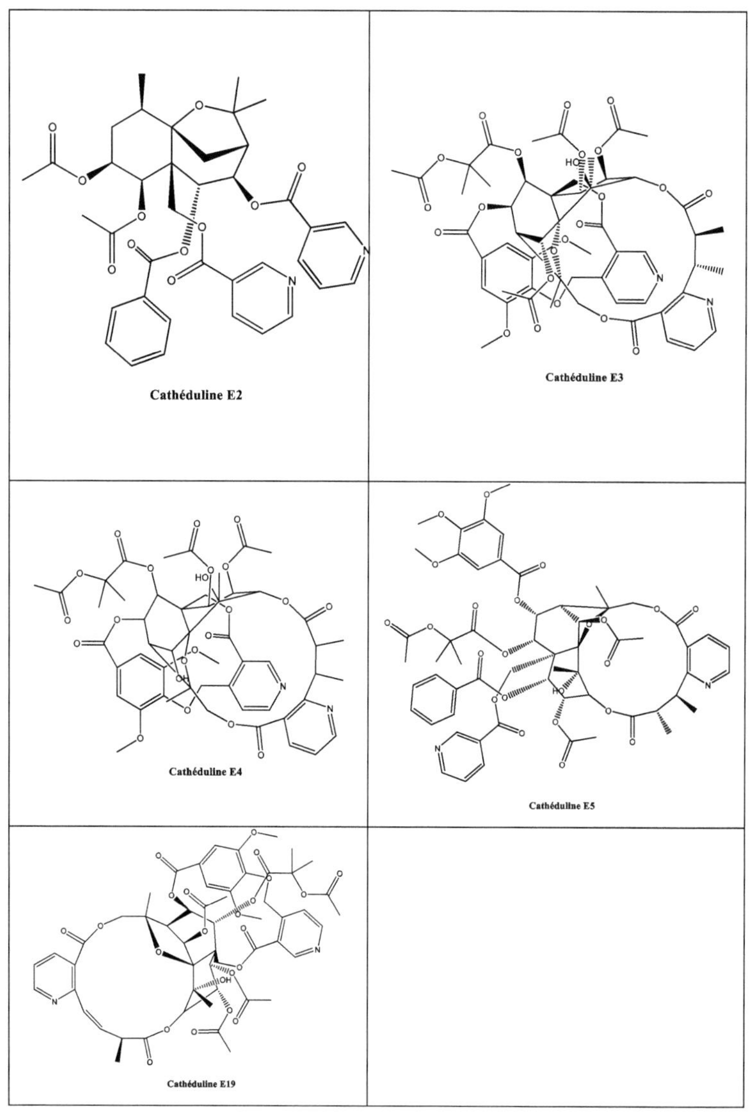

Medium molecular weight cathedulins :

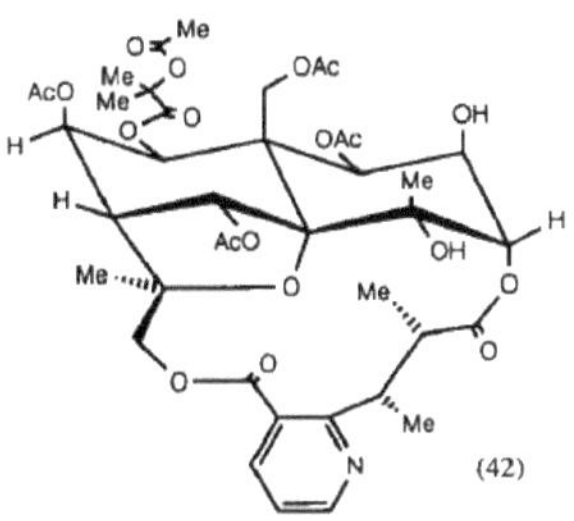 (42)	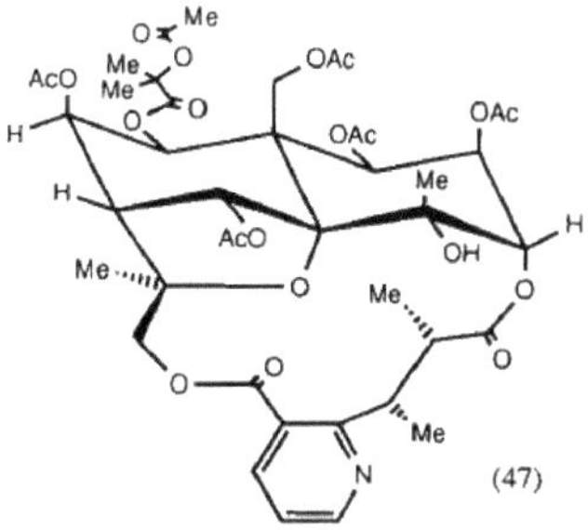 (47)
Cathedulin K2	Cathedulin K6
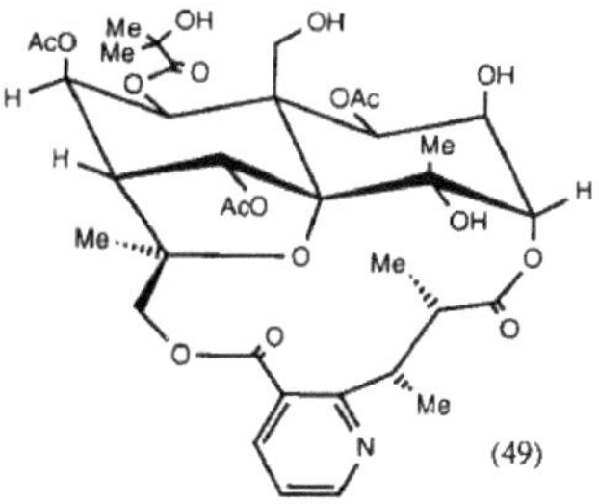 (49)	
Cathedulin K15	

High molecular weight cathedulins :

(50)	(51)
Cathedulin E3	Cathedulin E4
(57)	(58)
Cathedulin E5	Cathedulin E6
(59)	(60)
Cathedulin k12	Cathedulin k20

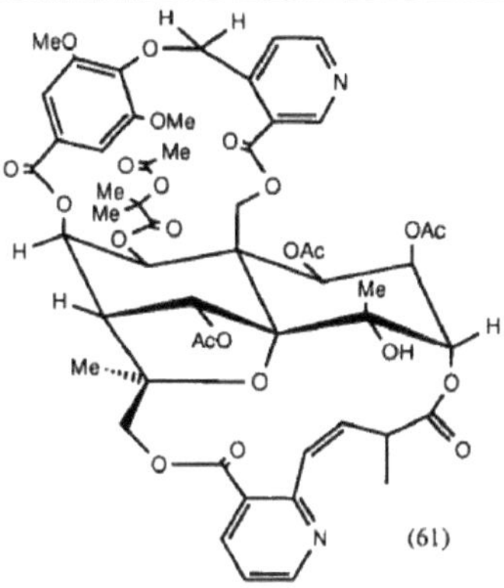

Cathedulin k19	Cathedulin k17

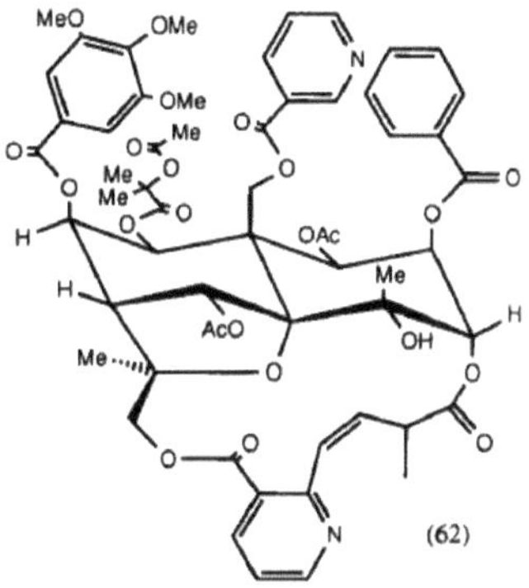

Appendix 6: Triterpene quinones.

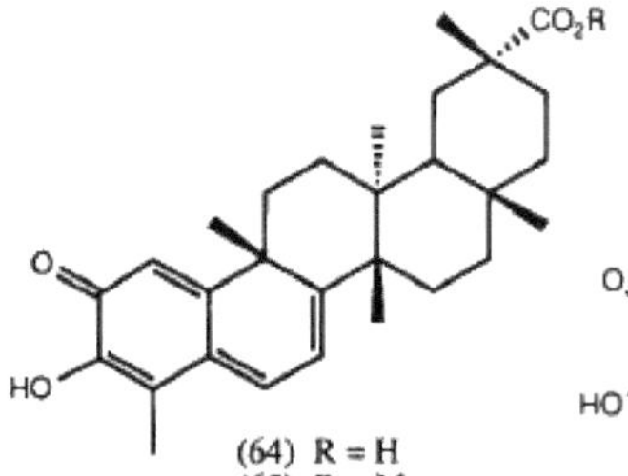

(64) R = H
(65) R = Me

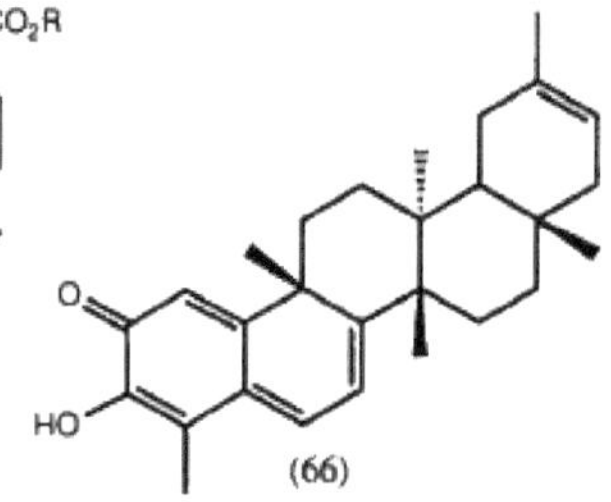

(66)

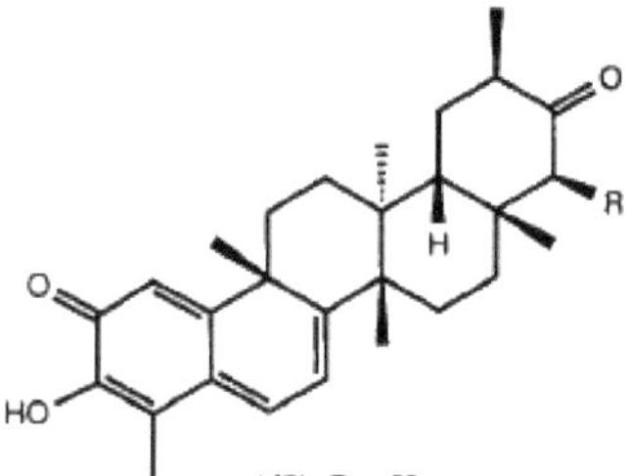

(67) R = H
(68) R = OH

64. Celastrol 65. Pristimerine 66. Iguesterine 67. Tingenin A 68. Tingenin B

Printed by Books on Demand GmbH, Norderstedt / Germany